Dedicated to the memory of
***M. J. E. Golay** (1902–1989),*
the inventor of open-tubular
gas-chromatographic columns.

Acknowledgements

Many of the figures in this book appear with the permission of the copyright holders, and these are referenced where they appear. Other figures and tables that have no references are from the collection of the authors and their associates, whose contributions are gratefully acknowledged.

Contents

Foreword   ix

PART I: Introduction   1

    1.1   The Beginnings   3
    1.2   Nomenclature   5
    1.3   Evolution of Open-Tubular Columns   6
            Column Material
            Sample Introduction
            Variation of Column Diameter and Film Thickness
            Improvements in the Stationary Phases
            Possible Future Developments
    1.4   The Chromatographic Process   10
    1.5   Types of Chromatographic Separation   13

Part II: The Carrier Gas   15

    2.1   Pressure Drop   16
    2.2   Linear Velocity   18
    2.3   Flow Rate   21
    2.4   Compression Correction Factor   22
    2.5   Carrier Gas Viscosity   24

Part III: Separation   27

    3.1   The Distribution Process   27
            Distribution Constant

Retention Factor
Phase Ratio
Effect of Stationary-Phase Film Thickness
3.2 Measuring Separation   32
Separation Factor
Relative Retention
3.3 The Retention Index System   35
Isothermal Retention Index
Programmed-Temperature Retention Index
Other Retention Index Systems

Part IV: Band Broadening and Resolution   41

4.1 Band Broadening   41
Effects of Band Broadening
Measurement of Peak Width
4.2 The Theoretical Plate Concept   46
Calculating the Number of Theoretical Plates
Effective Plate Number
4.3 The Origins of Band Broadening   50
The Minimum Theoretical Plate Height
The B and C Term Contributions
The C Term Without the Stationary-Phase Contribution
The C Term Including the Stationary-Phase Contribution
Influence of Retention on the Theoretical Plate Height
Percent Utilization of Theoretical Efficiency
4.4 Separation Quality   61
Resolution
Peak Number
Separation Number (Trennzahl)

Part V: The Stationary Phase   67

5.1 The Role of the Stationary Phase   67
Solute–Stationary Phase Interaction
Stationary-Phase Composition
Stationary-Phase Maximum Temperatures
5.2 Open-Tubular Column Types   76
Wall-Coated (WCOT)
Support-Coated (SCOT) and Porous-Layer (PLOT)

5.3	Column Materials	78
	Metal	
	Glass	
	Fused Silica	
5.4	Column Coating	82
	Dynamic Coating	
	Static Coating	
	Cross Linking and Immobilization	
5.5	Measuring Column Quality	86
5.6	Selecting a Stationary Phase	88

Part VI: The Variables of Open-Tubular GC Columns   91

6.1	Comparison to Packed Columns	92
6.2	Relationship of Efficiency, Selectivity, and Retention	96
6.3	Column Variables	97
	Column Inner Diameter	
	Column Length	
	Stationary-Phase Film Thickness	
	Influence of Column Variables on Sample Capacity	
6.4	Carrier Gas Variables	111
	Carrier Gas Velocity	
	Carrier Gas Type	
6.5	Speed of Analysis	115
6.6	The Effect of Column Temperature	119
	Isothermal Elution	
	Programmed-Temperature Elution	
	Temperature Program Optimization	

Part VII: Inlets   127

7.1	The Sample Introduction Process	127
	Column Requirements	
	Extra-Column Contributions to Band Broadening	
	Inlet Requirements	
7.2	Split Inlets	131
	The Split Ratio	
	Split Injection Linearity	
	Liner Packing	
	Adsorption and Decomposition	

7.3	Splitless Injection	137
	Recovering Peak Shape	
	The Solvent Effect	
	Use of a Retention Gap	
7.4	Direct Inlets	140
7.5	On-Column Inlets	142
7.6	Programmed-Temperature Injection	144
7.7	Selecting an Inlet	147

Part VIII: Detectors   149

8.1	Detector Types	150
8.2	Detector Requirements for Open-Tubular Columns	150
	Volume and Time Constant	
8.3	Flame-Ionization Detector	152
8.4	Thermal-Conductivity Detector	153
8.5	Electron-Capture Detector	155
8.6	Other Selective Detectors	157
	Mass-Spectrometric Detector	
	Nitrogen-Phophorus Detector	
	Photoionization Detector	
	Electrolytic-Conductivity Detector	
	Flame-Photometric Detector	
	Tandem Detection	

Part IX: Literature on Open-Tubular Columns   163

9.1	References	163
9.2	General References	170
	Textbooks	
	Chromatography Journals	
	Symposia Proceedings	
	Publishers	

Part X: List of Symbols   177

10.1	List of Symbols Used	177
	Symbols	
	Greek-Letter Symbols	
10.2	Acronyms and Abbreviations	180

Index   183

Foreword

"Open-Tubular Columns — An Introduction"* was originally published in 1973 by the Perkin-Elmer Corporation, followed five years later by a revised edition**. The first two editions of this short introductory text were very successsful; they were adopted by a number of open-tubular GC short courses, translated into several other languages, and continued to be in demand even through 1993.

In the past twenty years, open-tubular column gas chromatography has advanced tremendously in the areas of injectors, columns, and detectors. Many technologies that were nearly unheard of in the 1970s now occupy an important position in day-to-day laboratory practice; certainly, the ubiquitous flexible fused-silica columns are the most obvious example. Programmed-temperature injection, bonded stationary phases, and detectors specially designed for open-tubular col-

---

* L.S. Ettre, *Open-Tubular Columns – An Introduction*. Perkin-Elmer, Norwalk, CT, 1973, 64 pp.
** L.S. Ettre, *Introduction to Open-Tubular Columns*. Perkin-Elmer, Norwalk, CT, 1979, 74 pp.

umns also figure prominently among the technical developments of this period.

But perhaps most significant is the extent to which open-tubular columns have permeated the field and applications of gas chromatography. One only has to look to the published literature to realize the extent of this expansion. The series of International Symposia on Capillary Chromatography, running from 1975 to the present, presents an excellent chronology of the recent two decades (a summary of the symposia appears in Part IX of this book). We have not attempted to include a comprehensive bibliography; instead, we present selected references that pertain to specific topics. The reader is encouraged to explore the listed articles and texts for further study.

On the twentieth anniversay of the original edition, we are pleased to present this new book which considers all these advances in open-tubular column technology. This book was written primarily for the student and the day-to-day user of open-tubular gas chromatography. This edition includes a more detailed treatment of open-tubular GC theory. We have attempted to develop the theory in a logical and progressive manner, but of necessity we have not gone into great detail in many areas. The reader is encouraged to refer to the companion volume for further information on theoretical considerations*. We have included those topics that are essential for a practical and useful working knowledge of the subject. Some material on obsolete technologies has been removed, while areas dealing with the instrumentation have been expanded. These changes reflect the expansion of open-tubular gas chromatography from a specialized analytical technique to a routine analytical tool.

It is our hope that students as well as practitioners will find this book as useful as its predecessors.

April, 1994

John V. Hinshaw
Leslie S. Ettre

---

* L.S. Ettre and J.V. Hinshaw, *Basic Relationships of Gas Chromatography*, Advanstar, Cleveland OH, 1993. 190 pp. ISBN 0-929870-18-2.

PART I

Introduction

Open-tubular column gas chromatography is one of the few scientific disciplines founded on the basis of a theoretical insight which has developed into a major analytical technique promulgated across so many different application areas and engendering so many different ancillary techniques. The fundamental theory of open-tubular columns has, of course, been refined over the years; however, its basis remains undisturbed, and it is just as valid today as it was 35 years ago. In fact, the vast majority of work in open-tubular column research has concerned the investigation and refinement of techniques that could enable better usage of the theoretical potential of the column.

Also, although the theory was originally developed for gas chromatography, open-tubular columns are not unique in this respect: They have been employed in both liquid and supercritical-fluid chromatography. It is in GC, however, that the vast majority of open-tubular columns are applied.

It is interesting to note that practically all of the present methodology was already initiated at the beginning of the technique. Short as well as long columns, with smaller and

M. J. E. Golay (1902 – 1989). Photograph made a few months before his death.

larger diameters had been utilized; column tubing consisted of metal as well as glass. The 1959 patent of Desty et al.[1], describing a machine for the preparation of glass capillary tubing, already mentioned the possibility of using fused silica (quartz) as the tube material. Also, both thick- and thin-film columns, prepared using either the static or dynamic coating methods had been utilized. Furthermore, both split and splitless sample introduction techniques had their origins in the first years of open-tubular column gas chromatography. It was also recognized very early that small-mass capillary columns do not need the large thermostatted ovens used in commercial gas chromatographs; as early as 1962, a system was introduced (the Perkin-Elmer Model 226 gas chromatograph) in which the column temperature was controlled by conductive heating.

In this, the introductory chapter of our book, we shall briefly summarize the beginnings of open-tubular column gas chromatography, the circumstances surrounding the invention of these columns, and the first practical investigations demonstrating their superior performance. In addition, we shall highlight the most important steps in their further evolution. Details in this remarkable story (including further references) are documented in the gas chromatographic literature[2-5].

1.1 The Beginnings

Open-tubular (capillary) columns were invented by M.J.E. Golay*. He became involved in gas chromatography in 1955 after joining the Perkin-Elmer Corporation as a scientific advisor. The first commercial gas chromatograph had just been introduced: Particular interest was centered on the separation process occurring in a (packed) column and how they were related to column characteristics and the operation parameters. From theoretical considerations, corroborated by experiments, Golay concluded that in a packed column there is very little control over the geometry of the packing material and the path of the sample molecules during their passage through the column. Therefore, Golay decided to study a simplified system: a capillary tube in which the stationary phase is coated on its inner surface and the sample molecules have a straight, open, unrestricted path to travel.

The first experiments were carried out in the fall of 1956, clearly indicating the advantages of such a system. This was

---

* For a detailed discussion of the events related to the invention and early development of open-tubular columns by Golay, see Ettre[6].

then followed by systematic investigations and the development of the theory of these columns. Thermal conductivity detectors (TCD) used at that time had an inherently too-large volume for the new type of columns, through which the flow was about one-twentieth of packed-column flow rates. Therefore, Golay designed and built a remarkable small-volume, micro-TCD which enabled the systematic early development.

The first public report on the new type of columns and a crude theory were presented by Golay in August 1957, at the Symposium on Gas Chromatography organized by the Instrument Society of America[7]. Finally, a full discussion of the theory of open-tubular columns — together with the presentation of real chromatograms — was given by Golay in May 1958, at the Second International Symposium on Gas Chromatography organized by the British Gas Chromatography Discussion Group, in Amsterdam, The Netherlands[8].

The early development of open-tubular columns by Golay coincided with the development of the flame-ionization detector (FID) by McWilliam and Dewar of I.C.I. Australia[9]. A detailed discussion of the FID at the Amsterdam Symposium[10] resulted in the immediate realization that the FID was the ideal detector for these columns. Thus, this detector was already utilized in the first commercial GC employing Golay's columns, which was introduced in March 1959 by Perkin-Elmer[11]. The FID was also incorporated by Desty, at British Petroleum, into his breadboard system employing capillary columns that he first described in October 1958[12]. Another ionization detector introduced in 1958 was Lovelock's argon-ionization detector[13]. Its micro version[14] was used by Zlatkis[15] and Lipsky[16] in their publications on the use of open-tubular columns, the first utilization of these columns in the United States outside Golay's group.

Thus, by the spring of 1959, one year after the publication of the theory of open-tubular (capillary) columns and the first indication of their superior separation power, their application had been demonstrated by a number of independent groups. From then on, the evolution was straightforward.

1.2 Nomenclature

In his first paper dealing with the use of these columns[7], Golay used the expression "capillary" to emphasize the small diameter of the columns (0.25–0.50 mm) investigated by him as compared to the packed columns (about 4 mm) used at that time. Although the theoretical advantage of these columns is not related to their "capillary" dimension, the name stuck, in spite of the fact that — realizing this error — Golay was very careful in using the term *open-tubular column* in his 1958 Amsterdam paper[8]. In 1960, he again came back to this question, stating that while most open-tubular columns used in practice have a fairly small diameter, "it is not the smallness, it is the 'openness' of the open-tubular column which permits us to realize a two-orders of magnitude improvement over the packed column." In order to further emphasize this, he added that in fact "the 24-inch diameter gas pipe with an oil-coated inner wall, stretching from Texas to Maine" would indeed be an excellent open-tubular column for the analysis of high-boiling hydrocarbons[17].

The ambiguity of the "capillary column" term is also recognized in the most recent unified Nomenclature For Chromatography issued by the I.U.P.A.C.[18] which states that such a column may either contain a packing *(packed capillary column)* or have the stationary phase, consisting of either a liquid or an active solid, supported on its inside wall *(open-tubular column)*. Furthermore, the I.U.P.A.C. nomenclature emphasizes that open-tubular columns are characterized by the fact that "there is an open, unrestricted path for the mobile phase."

Thus, the correct expression for the columns developed by Golay and discussed in our book is *open-tubular columns*. As a colloquialism, "capillary column" may be tolerated, but one must be careful not to propound absurdity with expressions such as "large-diameter capillary columns." This clear distinction in the name used for these columns is not only a question of semantics. *Packed* "capillary" columns are used in both gas and liquid chromatography, and when this term is used, the reader is often confused as to which type of column is being mentioned.

1.3 Evolution of Open-Tubular Columns

A brief summary of some of the important events in the evolution of open-tubular columns is given below.

1.3.1 Column Material

Golay used glass and plastic tubing in the first experiments but soon switched to metal (mainly stainless steel) tubing, and these were used overwhelmingly in the first decade of open-tubular columns by practically everybody. Stainless steel was used because methods for the preparation of a stable coating on the inside of a *glass* capillary tube had not yet been developed. This happened only in the latter part of the 1960s, due mainly to the work of K. Grob, in Switzerland, and to Novotny and Tesarik, in Czechoslovakia. Because of their efforts, glass slowly replaced the metal columns around 1970. This transition was accelerated by the availability of laboratory machines permitting the drawing of glass capillary tubes of various diameters. These machines were all based on the system described by Desty and co-workers[1, 19]. As mentioned earlier, Desty raised the possibility of preparing capillary tubing from *quartz (fused silica)* and in 1975, he even showed how the glass drawing machine could be modified to make columns from quartz[20]. However, this was a fairly complicated system and the resulting tubing was rigid, with a fairly thick wall (just like the glass capillary columns). Dandeneau and Zerenner of Hewlett-Packard deserve credit for recognizing the advantages of *flexible, thin-walled* fused-silica tubing for capillary columns. Since their introduction in 1979[21], such columns have been used almost exclusively.

1.3.2 Sample Introduction*

Due to the reduced amount of stationary phase present in a capillary column — about 3–5 % of the amount present in a standard packed column — only a small fraction of the sample amount introduced into a packed column may be introduced into a capillary column. Even regular microsyringes are too large for the direct introduction of such small amounts. For this reason, the so-called *split injection* was developed almost at the beginning. The injected sample is evaporated, mixed with the carrier gas, and then the flow is split into two highly unequal portions, the smaller one being introduced into the column and the remainder discarded[11, 12]. Since then, split sample introduction has been continuously improved, and it remains the most widely used sample introduction mode.

However, even in its most advanced design, split sample introduction has serious shortcomings when trace-level samples are analyzed since most of the sample does not enter the column. In such a case, *non-split systems* are useful. The whole sample is deposited on the column, and additional techniques are employed to recover peaks shapes and control the side effects of large solvent volumes.

1.3.3 Variation of Column Diameter and Film Thickness

From their beginnings, the most frequently used open-tubular columns had an internal diameter of about 0.25 mm (0.010 in.); however, by 1959, columns having an internal diameter of about 0.51 mm (0.020 in.) had also been introduced commercially. These diameters remain the most widely used, even after the introduction of fused-silica tubing as the column material**.

---

* For a detailed discussion of the questions associated with sample introduction into open-tubular columns, see the books of K. Grob and of Sandra, listed in Section 9.2.1.
** The present-day 0.53–mm internal diameter was chosen to permit on-column injection of samples with standard microsyringes of 0.47–mm o.d. (see Part VII).

Columns with an internal diameter larger than 0.5 mm were also used almost from the onset, mainly to increase the permissible sample size[22] or for the analysis of high-boiling compounds. For example, Quiram demonstrated in 1963 that, using a 76–m long, 1.65–mm i.d. column, with a film thickness of about 0.8 μm, cetyl alcohol (hexadecanol, boiling point 344 °C) eluted in eight minutes at a column temperature of 175 °C (i.e. 169 °C below its boiling point)[23].

Adjustment of the tube diameter is usually related to a change in the thickness of the stationary (liquid) film coated on the inside wall of the column tubing. At the beginning, the selection of film thickness had to be compromised. A too-thin film resulted in poor chromatograms due to the activity of the (metal) tube's surface which interacted with the sample through the thin film, particularly when analyzing polar samples on non-polar phases. At the same time, too-thick films were not stable: They soon lost most of their coating through evaporation ("bleeding"). These problems were finally solved by the introduction of more inert tube material (glass, fused silica), and by immobilization (bonding) of the stationary phase film (see the next section). Thus, through the proper adjustment of both the tube diameter (from about 0.1 to 1.5 mm) and the film thickness (from about 0.1 to 5 μm), a wide variety of open-tubular columns can be prepared that are suitable for most applications.

1.3.4 Improvements in the Stationary Phases

The stationary (liquid) phases used at first were mostly low-molecular-weight compounds — squalane, phthalates, sebacates — with fairly high vapor pressures. Even the polymers utilized at that time were relatively small molecules; for example, the then widely-used poly(ethylene glycol), Carbowax 1540, had an average molecular weight of only 1540. These materials were not prepared for GC; chemicals readily available in the laboratory were utilized. This fact seriously limited the upper working temperature. Even in 1975, one could rarely find a capillary chromatogram in which the column was heated above 200 °C. This situation changed drastically in the 1970s by the introduction of silicone phases containing methyl, phenyl and cyanopropyl groups, having high molecular weights, synthesized specially for chromatography. This was then followed by the development of procedures to repolymerize a silicone phase *in situ* in the column, cross-linking and chemically bonding it to the inner surfaces of glass and fused-silica tubing. Open-tubular columns which can be used at high temperatures, up to and even above 400 °C, could be prepared by these procedures. In addition, the immobilization of the stationary (liquid) phases permitted the preparation of stable columns with almost any film thickness.

A further development has been the preparation of open-tubular columns with a porous solid adsorbent layer coated (prepared) on the inside tube wall. These columns take advantage of their "openness" in the analysis of gases which, up to then, could only be analyzed on packed columns containing adsorbent-type stationary phases.

1.3.5 Possible Future Developments

The evolution of open-tubular column gas chromatography is by no means finished; it is still a continuing process and future improvements can be expected. These improvements will probably be related mainly to miniaturization: reducing the diameter and length of the columns, with a concomitant reduction in the film thickness to maintain the phase ratio. Such a change, however, would significantly decrease the sample size that can be introduced, creating a problem in the instrumentation. This problem could be partly compensated by modifying the inside surface of the column tubing, thereby increasing the coated surface area. Still, new inlet systems and detectors will have to be designed which are suitable for the smaller sample sizes. These developments, however, will not change in the theory and fundamental relationships of open-tubular columns discussed in our book.

1.4 The Chromatographic Process

Chromatography is a physical separation process in which two or more analytes, or *solutes*, are distributed between a *stationary phase* and a *mobile phase* which moves past the stationary phase in a defined direction[24]. Among many variants, chromatographic separation can take place inside a tube which holds the stationary phase and conducts the mobile phase past it. This tube is called a *column*. The column has certain physical environmental requirements for proper operation which are maintained by devices surrounding and connected to it. Collectively these devices are called a *chromatograph* (see Figure 1).

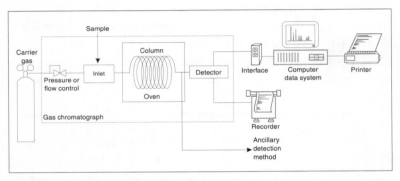

Figure 1. *Simplified schematic of a gas chromatograph.*

The column is suspended inside a temperature-controlled environment, usually an air bath *oven*. The oven temperature is closely controlled at a constant temperature for *isothermal* operation, or it is increased at a constant rate from an initial temperature to a final temperature for *temperature-programmed* operation. The column entrance is connected to a supply of mobile phase with an inlet device that introduces a small portion of the sample analyte mixture into the mobile phase stream. The sample, mixed with mobile phase, is swept into and through the column. Individual solutes pass through the column at different rates according to their physico-chemical

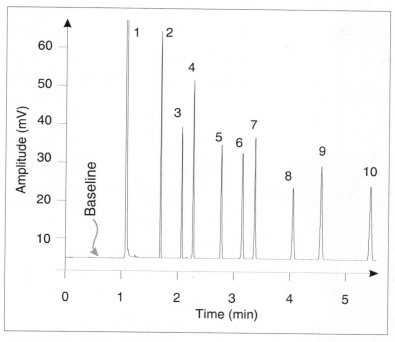

Figure 2. Chromatogram of a column test mixture. **Peak identification:** 1 = Solvent (isopropanol); 2 = n-nonane; 3 = 2-octanone; 4 = n-decane; 5 = 1-octanol; 6 = 2,6-dimethylphenol; 7 = n-undecane; 8 = 2,4-dimethylaniline; 9 = naphthalene; 10 = n-dodecane. **Column:** 25-meter x 0.25-mm i.d. x 0.25-μm film, methylsilicone phase on fused silica. **Conditions:** column oven, 100 °C; helium carrier gas; split injection; flame ionization detector.

relationships with the stationary phase and emerge from the column exit at distinct times after their injection.

The column exit is connected to a device, called a *detector*, that responds to an analyte with an electrical signal proportional to the amount of analyte present. The signal is amplified and plotted *vs.* time elapsed since sample injection. This recorder trace, together with amplitude and time information, is called a *chromatogram* (see Figure 2). Individual analytes produce signal deviations from zero that are called *peaks*, labeled 1–10 in Figure 2. The flat zones between the peaks are called the *baseline*. Each peak corresponds to one or more analytes: All may not completely separate during their passage through the col-

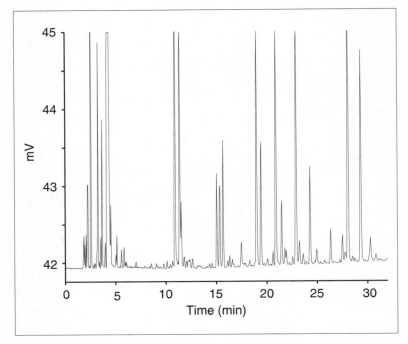

Figure 3. Separation of orange oil. **Column:** *25-m x 0.25-mm i.d. x 0.25-μm film, Carbowax-20M on fused silica.* **Conditions:** *Oven, 80 °C, isothermal for 5 min, then 2.5 °C/min to 220 °C. 0.2-μL split injection.*

umn. In most modern chromatographic systems, the detector signal is fed to a computerized *data-handling system* where it is analyzed and stored in a digital format. A chart recorder is sometimes used to record chromatograms, as well.

A chromatogram contains a wealth of information about the original sample, the chromatograph, and the column itself. It may be relatively simple, or much more complex as shown, for example, in Figure 3. The probable identity and quantity of individual sample components are the two most significant data in a chromatogram. A good understanding of the chromatographic process itself is important before a full and accurate appreciation of a chromatogram can be obtained.

As we investigate the chromatographic process in more detail, remember that while each analyte has a defined position in the chromatogram, there may be other analytes with the same position. Thus, the appearance of a peak at a known position does not necessarily mean that a particular analyte was present in the original sample, nor does it mean that the peak represents only one substance. On the other hand, the absence of a specific peak indicates that the corresponding substance was not present in the sample.

1.5 Types of Chromatographic Separation

The physical state of the mobile phase distinguishes the fundamental type of a chromatographic separation. Liquid chromatography (LC), gas chromatography (GC), and supercritical-fluid chromatography (SFC) are all named for the state of their respective mobile phases. Each of these mobile phases engenders unique separation characteristics that can be applied to different analytical or preparative problems. This book is concerned only with analytical GC separations on open-tubular columns. The interested reader is directed to the companion book by Ettre and Hinshaw[24], which gives a more detailed comparison of GC, LC, and SFC, as well as a comprehensive treatment of the principal theoretical relationships of gas chromatography on both packed and open-tubular columns.

Another distinguishing feature of the chromatographic techniques is the way in which the stationary phase is held inside the column. In packed-column chromatography, the stationary phase is coated onto small particles that are packed into the column bore, around which the mobile phase flows. Uncoated particles themselves may constitute the entire stationary phase in the case of porous polymers, molecular sieves, or gas-solid chromatographic adsorbents. In open-tubular column gas chromatography, the separation occurs in an open tube on which the stationary phase is coated, immobilized, or otherwise suspended (see Figure 4 on page 16).

In addition to packed columns, there are three principal ways in which stationary phases are held inside open-tubular columns (see Figure 25 on page 77). In *wall-coated open-tubular*

(WCOT) columns, the stationary phase is coated uniformly as a thin film on the column inner wall. In addition, the stationary phase may be chemically *bonded* to the column inner wall, and/or it may be extensively *cross-linked* or *immobilized* by self-polymerization after deposition into the column. The stationary phase may be dispersed on inert particles which adhere to the column wall, producing a *support-coated open-tubular* (SCOT) column. Or, the column inner wall may be etched, creating a porous surface onto which the stationary phase is deposited. This porous layer may also consist of an adsorbent acting as the stationary phase. Such columns are called porous-layer open-tubular (PLOT) *columns*.

✳ ✳ ✳

Open-tubular column chromatography can give much higher separation speed and efficiency compared to packed-column operation. Particulate column packings that obstruct the column bore significantly reduce the resolving power and speed of separation compared to open-tubular columns. The inherent openness of open-tubular columns permits the practical operation of longer columns, thereby generating more efficient separations. We present a comparison of packed and open-tubular columns in Part VI.

Part II

The Carrier Gas

In this chapter, we will explore the nature of carrier gas flow through open-tubular columns. In gas chromatography the mobile phase, or *carrier gas*, is an inert gas such as helium, hydrogen or nitrogen, which is applied under pressure to the column inlet. In other chromatographic separations (LC, SFC), the mobile phase is not always inert but possesses specific solvation properties that enter into the separation process. In GC, however, the mobile phase acts as a truly inert "carrier" that transports analyte molecules down the column, engaging only in non-specific interactions affecting all solutes equally*. Figure 4 on the next page illustrates many of the quantities we will discuss in Part II.

---

* Theoretically, it may be possible to use a mobile phase in GC that is not completely inert but has some interaction with the sample components, the stationary phase, or the support particles. An example is the use of steam as the carrier gas. Except for a few limited investigations, however, such systems have not been used in practice.

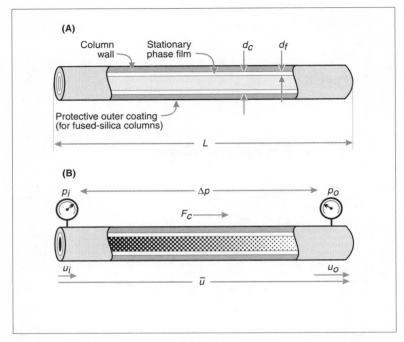

Figure 4. Open-tubular columns. (A) Column dimensions. (B) Column flows and pressures. Relative dimensions are exaggerated for emphasis.

2.1 Pressure Drop

In open-tubular GC, carrier gas is applied under pressure to the column inlet. The flow of carrier gas through the column is determined by the column *inner diameter*, d_c (mm*) and *length*, L (m), in relation to the applied carrier gas pressure drop and the carrier gas viscosity (see Section 2.5) at the column temperature. Figure 4(A) illustrates the fundamental dimensions of an open-tubular column. The column dimensions are exaggerated in this figure. In particular, note that the stationary phase *film thickness*, d_f (μm), is normally so thin in relation to the column

---

* Commonly-used metric units are given for each term in this book where first mentioned.

tube i.d. that its effect on the coated column i.d., and thus the carrier gas flow, can be safely ignored.

As carrier gas flows through the column, its pressure drops from the *inlet pressure*, p_i (kPa), to the *outlet pressure*, p_o (kPa) which is usually ambient or "atmospheric" pressure* (see Figure 4(B) for an example). There are two important quantities related to the *absolute* carrier gas pressures at the column inlet and outlet. First the *pressure drop*, Δp (kPa), across the column is the difference between the absolute inlet and outlet pressures:

$$\Delta p = p_i - p_o \qquad \text{eq.2.1}$$

This is the "gauge" or readout pressure shown on a gas chromatograph when the column outlet is at atmospheric pressure. It is important not to confuse the "gauge" pressure drop with the absolute inlet pressure. The former is most often reported as the "inlet pressure," while the latter is rarely noted in everyday use. When using pounds-per-square-inch, we have designated pressure differentials (pressure drops) with the units *psig* and absolute pressures with the units *psi*.

The second quantity is the *relative pressure* across the column, P, which is the *ratio* of the absolute inlet and outlet pressures:

$$P = \frac{p_i}{p_o} \qquad \text{eq.2.2}$$

The relative pressure is used in calculations involving the compressibility of the carrier gas (see section 2.4 on page 22).

---

* Vacuum outlet pressures are encountered with mass-spectrometric detection. See Section 8.6.1 for more information.

	psi			kPa	
Δp	p_i	P	Δp	p_i	P
3.63	18.1	1.25	25.0	125.0	1.25
7.25	21.8	1.50	50.0	150.0	1.50
14.5	29.0	2.00	100.0	200.0	2.00
36.3	50.8	3.50	250.0	350.0	3.50
72.5	87.0	6.00	500.0	600.0	6.00

Table 1. Pressures and pressure drops in pounds-per-square-inch and kilopascals. p_o = 14.5038 psi or 100 kPa (1 bar).

Carrier gas pressures are most often expressed in units of kilopascals (kPa) or pounds-per-square-inch (psi). To convert to psi, multiply the pressure in kPa by 0.145038. To convert to kPa, multiply the pressure in psi by 6.89476:

kPa = 6.89476 x psi
psi = 0.145038 x kPa

Table 1 gives examples of absolute and relative pressures in both kPa and psi.

2.2 Linear Velocity

Carrier gas moves through a column at certain speeds or *linear velocities*, u (cm/s) (see Figure 4(B) on page 16). The linear velocity affects both the quality and speed of a chromatographic separation. It is necessary to determine the velocity for any particular separation so that the analysis can be replicated at a later time, and the separation can be optimized for speed and efficiency.

As the carrier gas moves from the beginning of the column to the column exit, its linear velocity increases from that at the column inlet, u_i (cm/s), to the carrier gas velocity at the column outlet, u_o (cm/s). This behavior is due to the compressible nature of the carrier gas which expands as it moves through the column (see Section 2.4). The *average linear velocity*, $\bar{u}$ (cm/s) at which the carrier gas transits the column is customarily used to express the overall velocity through a column. It can be calculated from the interval that a solute spending all of its time in the mobile phase (an *unretained solute*) takes to move from the beginning to the end of the column. Its elution time is the *unretained peak time* (gas holdup time), t_M (s). Methane has been used frequently for this purpose*.

The average linear velocity of the carrier gas through the column is calculated from the unretained peak time and the column length, L (cm):

$$\bar{u} = \frac{L}{t_M} \qquad \text{eq.2.3}$$

This is the average value across the entire column length. A value of between 20 and 30 cm/s is typically set at the column operating temperature. This is close to the optimum velocity for most open-tubular columns with helium carrier gas.

Setting the average linear velocity involves measuring the unretained peak time, calculating the velocity, and then adjusting the inlet pressure until the desired velocity is reached. This repetitive process can be shortened by estimating the required pressure drop at the desired velocity and using the estimate for the first velocity measurement.

---

* Methane is not always appropriate, since it is significantly retained on certain types of GC columns. In addition, the GC detector in use may not respond to methane, and another compound may be a better choice. The gas holdup time also may be calculated from the retention times of three members of a homologous series. For details, see Section 2.2 of ref. [24].

$\bar{u}$ (cm/s)	Column I.D. (d_c, mm)				
	0.10	0.20	0.25	0.32	0.53
	Pressure Drop (Δp, psig)				
20	52.8	13.2	8.5	5.2	1.9
30	79.2	19.8	12.7	7.7	2.8
40	106	26.4	16.9	10.3	3.8
50	123	33.0	21.1	12.9	4.7
60	158	39.6	25.3	15.5	5.6
70	185	46.2	29.6	18.1	6.6
80	211	52.8	33.8	20.6	7.5

Table 2. *Estimated pressure drops at various average linear velocities. $T_C = 100\ °C$, carrier gas = helium (viscosity = 2.28 x 10^{-5} Pa·s), column length = 25 m.*

The relationship between the pressure drop, Δp, and the average linear velocity, $\bar{u}$, is shown in eq.2.4:

$$\Delta p \approx \frac{32 \cdot L \cdot \eta \cdot \bar{u}}{d_c^2} \qquad \text{eq.2.4}$$

where L is the column length and η is the carrier gas viscosity (see Section 2.5). Eq.2.4 is accurate to within 25 percent*; the higher calculated pressures will be somewhat less than the actual pressures required to achieve a given velocity[25]. Values of Δp calculated from eq.2.4 are given in Table 2 for 25–m long columns with helium carrier gas at 100 °C. For 12.5–m columns, use half the pressure shown. For 50–m columns, double it.

---

* For a more accurate relationship, see Section 5.5 of ref. [24].

Example: Measuring and setting $\bar{u}$. A 25-meter long column with a 0.25-mm i.d. is installed in a gas chromatograph. The carrier gas inlet pressure drop (Δp) required for $\bar{u} = 30$ cm/s is estimated at 12.7 psig (from Table 2). The methane retention time, t_M, at this pressure is measured at 79 seconds. The average linear velocity is then calculated:

$$\bar{u} = (25 \times 100)/79 = 31.6 \text{ cm/s}$$

The factor of (100) converts the 25–m column length to centimeters, giving the result in cm/s. If $\bar{u}$ is not close enough to the desired value, the pressure is changed and the unretained peak time measured again. This process is repeated until the desired average velocity is attained.

The average linear velocity also plays a major role in determining an open-tubular column's performance or efficiency in separating the mixture under scrutiny (see section 4.3). We will examine the relationship of $\bar{u}$ to chromatographic performance after further discussion of the column and some of the basic properties of GC separations.

2.3 Flow Rate

Once the linear velocity has been set, the column exit flow rate at the column oven temperature (F_c) can be calculated. While not directly related to the column efficiency, the flow rate is important for both inlet and detector operation. Direct measurement usually is not used to determine F_c because open-tubular column flow rates can be low enough (down to about 0.2 mL/min) that physical measurement with a bubble flow meter is inaccurate*. A low-volume bubble flow meter can determine open-tubular column flows of about 2 mL/min and higher with sufficient accuracy, and is often used with larger

---

* Note that a bubble flow meter measures the flow at ambient (room) temperature (F_a) (see eq.2.9 on page 24). Correction for the water vapor pressure in a bubble flow meter is required; see Section 5.1 of ref. [24]. Flow meters that time the bubble electronically usually include this correction in their readouts.

diameter columns ($d_c \geq 0.32$ mm). Electronic flow meters are also available.

For all open-tubular columns F_c can be calculated from the carrier gas linear velocity at the column outlet, u_o, by multiplying it by the geometric cross-sectional area of the column ($\pi d_c^2/4$):

$$F_c = \frac{\pi d_c^2 u_o}{4} \qquad \text{eq.2.5}$$

The outlet linear velocity is determined by measuring the average linear velocity ($\bar{u}$), and then correcting to find u_o.

2.4 Compression Correction Factor

In open-tubular GC, the outlet linear velocity (u_o) is related to $\bar{u}$ by the *mobile-phase compression correction factor, j* (dimensionless):

$$\bar{u} = j\, u_o \qquad \text{eq.2.6}$$

The compression correction factor[26] expresses the way in which the carrier gas expands as it moves through the column. It is determined from the relative pressure across the column, P:

$$j = \frac{3}{2} \cdot \frac{P^2 - 1}{P^3 - 1} \qquad \text{eq.2.7}$$

Some typical values of j are given in Table 3.

With this information in hand, the column outlet flow rate can be expressed in terms of the directly measurable average linear velocity, column diameter, and relative pressure by combining equations 2.5 and 2.6:

$$F_c = \frac{\pi d_c^2\, \bar{u}}{4 j} \qquad \text{eq.2.8}$$

Δp (psi)	Δp (kPa)	P	j
0.0	0	1.00	—
7.3	50	1.50	0.789
14.5	100	2.00	0.643
21.8	150	2.50	0.538
29.0	200	3.00	0.462
36.3	250	3.50	0.403
43.5	300	4.00	0.357
50.8	350	4.50	0.320
58.0	400	5.00	0.290
65.3	450	5.50	0.265
72.5	500	6.00	0.244

Table 3. Values of j for different values of Δp and P.

Example: Calculating F_c. The average linear velocity at $\Delta p = 12.7$ psig was 31.6 cm/s on our 25–m x 0.25–mm i.d. column. The relative pressure, P, for this pressure drop is 1.88 (eq. 2.2), assuming that $p_o = 14.5$ psi:

$$P = (12.7 + 14.5)/14.5 = 1.88$$

The compression correction factor at this relative pressure is 0.675 (eq.2.7):

$$j = \frac{3 \times (1.88^2 - 1)}{2 \times (1.88^3 - 1)} = 0.675$$

Now we can calculate the column flow rate from eq.2.8:

$$F_c = \frac{\pi \times (0.25 \times 0.1)^2 \times 31.6}{4 \times 0.675} \times 60 = 1.49 \text{ mL/min}$$

Note that the column i.d. was converted to cm by multiplying the diameter in mm by 0.1. The resulting flow rate in mL/s was multiplied by 60 to convert to mL/min.

Thus, the determination of column outlet flow involves timing an unretained solute's passage through the column, determining $\bar{u}$, and applying eq.2.8.

The column outlet flow, F_c, is correct at the column oven temperature (T_c). For easy comparison to other flows measured at different temperatures, F_c is often further corrected to room temperature (T_a); the ratio of the absolute room temperature to the absolute column oven temperature is multiplied by the column flow:

$$F_a = F_c \cdot \frac{T_a}{T_c} \qquad \text{eq.2.9}$$

Here, the column and room temperatures are measured in degrees Kelvin.

> **Example: Correcting F_c to room temperature.** The calculated column outlet flow for the 25–m x 0.25–mm i.d. column at 100 °C was 1.49 mL/min. To correct this flow to a 25 °C room temperature, use eq.2.9:
>
> $$F_a = 1.49 \times \frac{25.0 + 273.15}{100.0 + 273.15} = 1.19 \text{ mL/min}$$
>
> The temperatures are corrected to degrees Kelvin by adding 273.15.

The column flow corrected to room temperature can be used, for example, to calculate the split ratio in an open-tubular column inlet system (see Section 7.2.1).

2.5 Carrier Gas Viscosity

The carrier gas viscosity affects column flow and linear velocity. Viscosity changes with temperature and is different for each type of carrier gas. Hydrogen has the lowest viscosity, and thus the lowest pressure drop for a given column length, inner diameter, and flow rate. Nitrogen is next highest, followed closely by helium. Figure 5 shows the relationship between

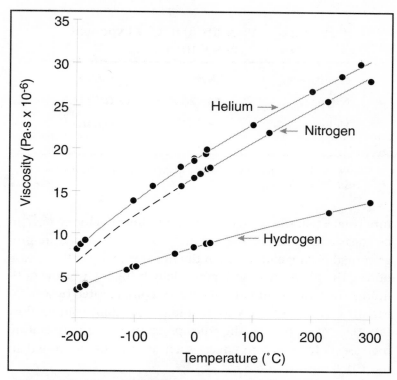

Figure 5. The effect of temperature on carrier gas viscosity. Points indicate experimentally-determined values. Lines are calculated according to eq.2.10 using the values in Table 4 [27].

column temperature and carrier gas viscosity; as the temperature increases the viscosity also increases, but in a nonlinear fashion. This function is given by eq.2.10:

$$\eta_i = \eta_0 \left(\frac{T_i}{T_0}\right)^x \qquad \text{eq.2.10}$$

where η_i is the viscosity at a temperature T_i (°K), η_0 is the known viscosity at a reference temperature T_0 (°K), and x is an exponent that depends on the carrier gas. Values of x and η_0 are given in Table 4 on the next page.

Viscosity effects are particularly important when the column oven temperature changes. If a new isothermal tempera-

Carrier gas	Viscosity at 0 °C (Pa·s x 10^{-6})	Exponent, x
Hydrogen	8.399	0.680
Nitrogen	16.736	0.725
Helium	18.662	0.646

Table 4. Viscosities at 0 °C and exponents for calculating viscosity at other temperatures.

ture is selected, the carrier gas flow and linear velocity have to be measured again. If the column oven is temperature-programmed during elution, then the linear velocity will decrease during the program period, potentially bringing it below optimum. The increase in viscosity can be compensated by arranging the carrier gas supply to maintain a constant column flow rate (i.e. by increasing the inlet pressure as the temperature increases). Temperature effects such as this are addressed in more detail in Section 6.6.

❋ ❋ ❋

The flow of carrier gas through an open-tubular column is governed by a set of well-defined relationships between the carrier gas and the column dimensions. The essence of chromatographic separation lies in the distribution between the solutes and the column's stationary and mobile phases. In Parts III and IV, we will present the basis of solute separation and resolution on open-tubular GC columns, and then in Part V we will return to look at the influence of the stationary phase in more detail.

Part III

Separation

Now that we have established the way in which the mobile phase flows through an open-tubular column, we introduce the processes that give rise to a column's ability to differentiate individual solutes.

3.1 The Distribution Process

Analytes passing through the column undergo discrete sorption/desorption steps at the molecular level between the mobile and stationary phases. This dynamic process is called *distribution (partitioning)*. During their residence in the column, analytes move along with the mobile phase while they are in it and are nearly motionless with respect to the column while in the stationary phase. Eventually, all analytes introduced into a column will come out of, or *elute* from the column exit (as long as the carrier gas continues to flow). All solutes spend equal time in the mobile phase if the average carrier gas velocity is constant. However, the time they spend in the stationary phase is different; those solutes less strongly attracted to the stationary

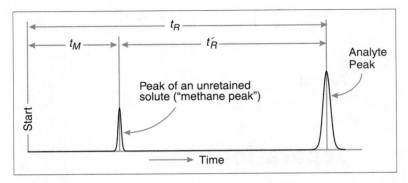

Figure 6. Retention times and related values measured from a chromatogram.

phase (*weakly retained*) spend less time in it and will elute first, while those more strongly attracted (*strongly retained*) will spend more time in the stationary phase and elute later. The details of solute–stationary phase interactions that give rise to these differences are discussed separately in Part V. In this chapter, we present the relationships that characterize the separation process.

The time that any solute spends moving with the carrier gas before elution from the column is equal to the unretained peak time (gas holdup time, t_M). The time that a solute spends distributed in the stationary phase is called the *adjusted retention time*, t_R' (s). The sum of the times spent in both the carrier gas and the liquid phase is equal to the *total retention time*, t_R (s), which is the elapsed time between solute introduction into the column and elution from the column exit:

$$t_R = t_M + t_R' \qquad \text{eq.3.1}$$

These relationships are illustrated in Figure 6. Since the average carrier gas linear velocity normally is held constant throughout an analysis, all solutes spend equal times traveling down the column in the mobile phase. Thus, it is only the differences in the time spent in the stationary phase that cause solutes to elute with different retention times.

3.1.1 Distribution Constant

As solutes move through the column, they are distributed (partitioned) between the mobile and stationary phases. If the distribution process is fast enough and the solute concentrations are low enough, then solute concentrations in both phases approach equilibrium. This quasi-equilibrium situation occurs for practical purposes in most GC analyses. The overall solute–stationary phase interaction can be expressed in terms of the *distribution constant (partition coefficient), K* (dimensionless), as the ratio of the equilibrium concentrations of a solute in the stationary and mobile phases during partitioning in the column:

$$K = \frac{c_S}{c_M} \qquad \text{eq.3.2}$$

where c_S and c_M are the equilibrium concentrations in the stationary and mobile phases, respectively. The distribution constant changes at different column temperatures or in different stationary phases; the larger the distribution constant, the stronger a solute's retention. Unlike the retention factor, however, the distribution constant remains the same even if the column's film thickness is changed. Thus, the distribution constant expresses the stationary phase's retentiveness towards a solute at a specific temperature.

The affinity of a solute for the stationary phase is controlled by two principal factors: the solute's vapor pressure (boiling point) and its activity coefficient in the stationary phase. Significant differences between either or both of these properties give rise to a column's ability to differentiate two solutes and elute them with different retention times. We will examine the influence of the stationary phase on retention in more detail in Part V.

3.1.2 Retention Factor

Solute retention is also expressed in terms of the *retention factor*, k (dimensionless), which quantifies the ratio of the time spent in the stationary phase to that spent in the mobile phase:

$$k = \frac{t_R'}{t_M} \qquad \text{eq.3.3}$$

$$k = \frac{t_R - t_M}{t_M} \qquad \text{eq.3.4}$$

The retention factor is independent of the carrier gas linear velocity. Higher or lower constant velocities or flow rates do not affect the relative times spent in the stationary and mobile phases, only their absolute values. Thus, k provides a convenient measure of retention without concern for the carrier gas flow. Changing the column temperature or the stationary phase type will affect the retention factor, however, because the relationship between the solute and the stationary phase has changed. Otherwise identical columns with different film thicknesses will also exhibit different k values (see the next section). Values of k are comparable only between solutes on a specific column under identical temperature conditions. In practical use, the retention factor is most often applied to calculations involving the column's efficiency (see Part IV).

3.1.3 Phase Ratio

The retention factor (k) is affected by the *amount* of stationary phase present. Larger stationary phase amounts will increase retention. For WCOT columns, the *stationary phase volume*, V_S (cm^3) may be calculated from the film thickness and the column's inner surface area ($\pi\, d_c\, L$):

$$V_S = (\pi\, d_c\, L)\, d_f \qquad \text{eq.3.5}$$

The ratio of the mobile phase volume (V_G) to the stationary phase volume is called the *phase ratio*, β (dimensionless). The

d_f (μm)	d_c (mm)				
	0.10	0.20	0.25	0.32	0.53
	Phase ratio, β				
0.1	250	500	625	800	1325
0.2	125	250	313	400	663
0.5	50.0	100	125	160	265
1.0	25.0	50.0	62.5	80.0	133
2.0	12.5	25.0	31.3	40.0	66.3
5.0	5.00	10.0	12.5	16.0	26.5

Table 5. Values of the phase ratio, β, for various column diameters and film thicknesses.

ratio is usually calculated neglecting the small reduction in column volume due to the stationary phase film (see Figure 4(A) on page 16):

$$\beta = \frac{V_G}{V_S} \approx \frac{d_c}{4 d_f} \qquad \text{eq.3.6}$$

As d_f increases, β will decrease. For commercially available open-tubular columns, β ranges from about 10 to over 500. Phase ratios for a number of typical open-tubular columns are given in Table 5.

3.1.4 Effect of Stationary Phase Film Thickness

The distribution constant (K) is related to the phase ratio (β) and the retention factor (k) by the following relationship:

$$K = \beta \cdot k \qquad \text{eq.3.7}$$

Thicker stationary phase films (smaller β values) will increase the retention factor, thereby producing longer retention times. In practice, the film thickness is selected so that specific solutes elute within a practical retention time range from about twice the unretained peak time ($k = 1$) to around 45 minutes. If all peaks do not elute within this time, then a thinner stationary phase film (higher β value) will reduce the retention times. On the other hand, if all the peaks elute too quickly they may not be completely separated, and a thicker stationary phase film may be appropriate. See Part VI for a more detailed discussion of open-tubular GC optimization.

3.2 Measuring Separation

A mixture of analytes is sent through the column so that they may be separated, detected, and measured. Differences in the solutes' physico-chemical properties in relation to the stationary phase and the column temperature cause them to separate and to elute from the column at different times. Figure 7 illustrates the situation that arises when two solutes have different retention times. Three peaks are shown. The first corresponds to an unretained peak (t_M), which may not always appear in the chromatogram. It is shown here for convenience as a reference point. Note that the retention factor of the unretained peak (k_M) always equals zero. A subscript is assigned to indicate the first and second solutes: The first elutes with a retention time, t_{R1}, and the second at t_{R2}. Each has an adjusted retention time, t_{R1}' and t_{R2}', and retention factor, k_1 and k_2. Since the column temperature must be constant for the k-values to be meaningful, these and the following measurements are valid only for isothermal operation. Programmed-temperature elution is considered in Section 6.6.2.

3.2.1 Separation Factor

The degree of separation of two adjacent solutes is expressed in terms of the *separation factor*, α (dimensionless), as

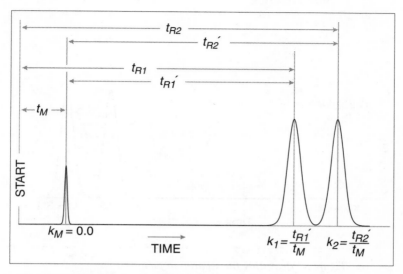

Figure 7. Retention time parameters measured from two solutes.

the ratio of the later eluting solute's retention factor to that of the earlier eluting solute:

$$\alpha = \frac{k_2}{k_1} \qquad \text{eq.3.8}$$

or as the ratio of their adjusted retention times:

$$\alpha = \frac{t_{R2}'}{t_{R1}'} \qquad \text{eq.3.9}$$

The separation factor is always greater than one (see Figure 8 on the next page). The larger the separation factor, the greater the difference in the solutes' retentions. A separation factor of exactly one corresponds to two *coeluting* solutes.

3.2.2 Relative Retention

The *relative retention*, r (dimensionless) expresses the degree of separation for two peaks that do not need to be adjacent. One of the solutes is designated as the *standard*, and the relative

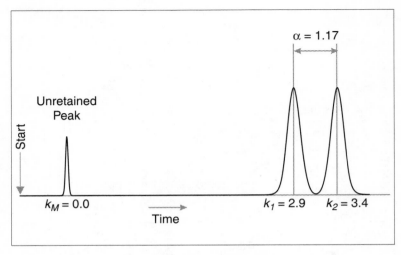

Figure 8. Measuring the separation factor, α.

retentions of one or more additional solutes are expressed in terms of the standard solute's retention factor:

$$r_i = \frac{k_i}{k_{st}}$$ eq.3.10

or in terms of the adjusted retention times:

$$r_i = \frac{t'_{R\,i}}{t'_{R\,(st)}}$$ eq.3.11

The symbol i refers to an individual solute, and the symbol (st) refers to the standard solute. Solutes eluting after the standard will have relative retentions greater than 1.0, while earlier solutes will have r values of less than 1.0. See Figure 9 for an illustration of the relative retentions of multiple solutes. The choice of a standard solute is arbitrary, but if it is next to peak i, and $k_{st} < k_i$, then $r_i \equiv \alpha$. Many chromatographic data handling systems use relative retentions to identify solutes, and may allow the choice of multiple retention-reference peaks. The accuracy of relative retention is degraded for peaks far away from the reference peak, but multiple reference peaks limit this

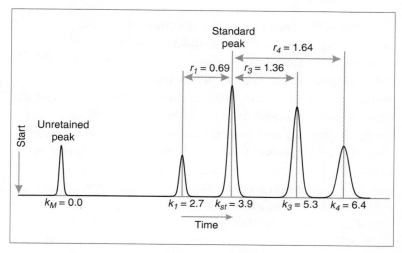

Figure 9. Relative retentions of multiple solutes.

problem by allowing each identified peak to be close to its reference point.

3.3 The Retention Index System

The retention time, the retention factor, and the relative retention all characterize peaks' retention behavior on a specific column. The absolute retention time changes with all of the chromatographic variables: linear velocity, temperature, phase ratio, and column length. Its utility as a qualitative identifier diminishes rapidly as normal run-to-run variations occur. Retention time is useless for identifying peaks on columns with different velocities, lengths, phase ratios, or temperatures. The retention factor is a little better; it accounts for some variations by inclusion of the unretained peak time, which encompasses velocity and length changes. Relative retentions include the phase ratio, as well, so that they can be compared on different columns, as long as the same reference peak is selected and the column temperatures are the same.

It was the lack of a suitable frame of reference that led Kováts to propose a *retention index* (*I*) system based on the homologous series of normal paraffins as reference peaks[28, 29]. The retention index system differs from relative retention in two ways. First, each analyte is referenced in terms of its position between the two *n*-paraffins that bracket its retention time. Thus, all analytes are close to their respective reference peaks. Second, the calculation is based on a linear interpolation of the carbon chain length ("carbon number") of the two bracketing normal paraffins. For convenience, the carbon numbers are multiplied by 100 to avoid the use of decimal fractions.

The retention index of a solute is equal to the carbon number (× 100) of a hypothetical *n*-paraffin that would have the same adjusted retention time as that specific solute.

Retention indices can be calculated from either isothermal or temperature-programmed data. Of these, the isothermal retention index is more accurate and reproducible.

3.3.1 Isothermal Retention Index

The original retention index system was based on isothermal GC. Under such conditions there is a semilogarithmic relationship between the adjusted retention times of the *n*-paraffins and their carbon numbers (c_n):

$$\log t_R' = a \cdot c_n + b \qquad \text{eq.3.12}$$

where *a* and *b* are constants. Figure 10 illustrates this relationship. By definition, the retention index of the *n*-paraffins is always equal to 100 times their carbon number. Thus, n-C_6 has an index of 600, n-C_7 has an index of 700, and so on. Generally, the relationship expressed by eq.3.12 is less accurate below *n*-pentane; thus, the same is true for retention index values below 500. There are, however, only a few organic compounds that fall in this region.

In Figure 10, the peak of analyte *i* (in this case, benzene) is located between the peaks of *n*-hexane and *n*-heptane; hence, its retention index lies between 600 and 700. The retention

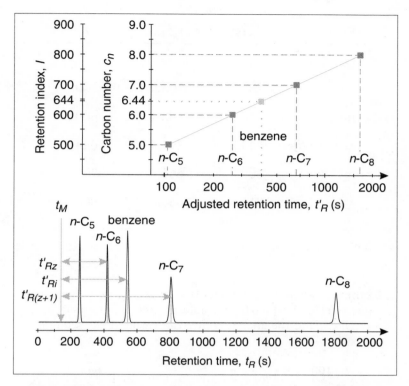

Figure 10. Calculation of the isothermal retention index. Solute i is benzene; its retention index is calculated as 644. Data refer to a 0.25-mm i.d. open-tubular column coated with methylsilicone phase and operated at 60 °C. See Table 6 on the next page for numerical data.

index can be calculated from the following equation with the respective adjusted retention time (t_R') values:

$$I = 100 \left(z + \frac{\log t'_{Ri} - \log t'_{Rz}}{\log t'_{R(z+1)} - \log t'_{Rz}} \right) \qquad \text{eq.3.13}$$

where we have assigned the carbon chain length of the first hydrocarbon to z, and designated the two bracketing hydrocarbons with subscripts z and $z+1$. The analyte is given the subscript i.

Peak	t_R (s)	t_R' (s)	I
t_M	147.9	—	—
n-C_5	251.8	103.9	500
n-C_6 (z)	410.0	262.1	600
benzene (i)	543.3	395.4	644
n-C_7 (z+1)	809.2	661.3	700
n-C_8	1816.8	1668.9	800

Table 6. Data for Figure 10.

Example: Calculating the retention index. Table 6 lists the data from Figure 10. Taking the adjusted retention times (t_R') for n-C_6, n-C_7, and the analyte (i), we can calculate the retention index, I:

$$I = 100 \times \left(6 + \frac{\log(395.4) - \log(262.1)}{\log(661.3) - \log(262.1)} \right) = 644$$

Retention index values are specific to one stationary phase at one temperature. In the case of a polar analyte and a nonpolar phase, any residual activity of the column tube's surface may influence retention of the analyte relative to the (nonpolar) n-paraffins. In such a case, the retention index will be influenced by the stationary phase film thickness; index values measured on a thinner film column tend to be higher than measured on a thicker film column since the residual surface activity will be shielded by the thicker film.

Retention index values are more accurate than single-reference relative retention values because the analyte and reference peaks are always within one carbon unit's elution time of each other. Retention indices make it possible to characterize analytes' behavior accurately on a stationary phase at a single temperature*. Some tables of isothermal retention indices have been published, although a number were based on measurements taken on stationary phases that are no longer used. Nonetheless, they are quite useful for establishing retention relationships among listed solutes.

Retention indices also have been applied to stationary phase characterization. The Rohrschneider-McReynolds system[30–32] uses differences in retention indices for various standard test probes on the phase under examination *versus* a nonpolar phase (squalane). A number of different test probes are employed to characterize a stationary phase's behavior towards different polar solute interactions. For example, benzene indicates aromatic ring interactions, while 1-butanol indicates hydroxyl and carbonyl interactions. This application of retention indices is detailed in Part IX of ref. [24].

3.3.2 Programmed-Temperature Retention Index

The retention index as calculated above refers only to data obtained under isothermal elution conditions. During column temperature programming, the series of *n*-paraffins elute in a linear fashion. Each successive peak adds a constant increment to the retention time of its predecessor, instead of an increasing increment as found with isothermal elution. Thus, the relationship of their *programmed-temperature retention times*, t_R^T, with the carbon number can be expressed as:

$$t_R^T = a' \cdot c_n + b' \qquad \text{eq.3.14}$$

where a' and b' are proportionality constants. This characteristic of programmed-temperature elution enables us to express the *programmed-temperature retention index*, I^T, in the following way[33]:

---

* The column oven temperature should be calibrated to within better than 1 °C for accurate retention index comparisons across more than one gas chromatograph.

$$I^T = 100\left(z + \frac{t_{Ri}^T - t_{Rz}^T}{t_{R(z+1)}^T - t_{Rz}^T}\right) \qquad \text{eq.3.15}$$

Retention index values calculated in this way are also called the *linear retention index* because of the linear nature of eq.3.14.

For a given solute and stationary phase, the values of the isothermal and programmed-temperature retention indices are not the same, although they are close. Values of I^T may also vary somewhat depending on the temperature programming rate and the initial temperature.

3.3.3 Other Retention Index Systems

A general theory of incremental structure–retention relationships predicts that any regular increment in a series of chemical structures should provide a regular increment in corresponding retention times. A number of other reference systems for retention indices take advantage of this behavior. Linear chain fatty acid methyl esters provide a reference system for finding the "equivalent chain length" of unsaturated and branched fatty esters. Another system uses the number of rings in unsubstituted polynuclear aromatic hydrocarbons (PAH) to characterize the retention behavior of their homologs and related substituted compounds.

※ ※ ※

In this chapter, we have discussed the fundamentals of solute separation on open-tubular GC columns. Although two peaks may be separated to some degree, they may not be completely resolved from each other: One peak may overlap the next. In the next part, we will introduce the concept of peak resolution and explore the origins of column effects that cause peaks to spread and overlap during their passage through the column.

Part IV

Band Broadening and Resolution

In Part III, we discussed the open-tubular GC separation of solutes based on differences in their chemical natures in terms of the chemistry of the stationary phase as the carrier gas flows through the column. However, even though the centers of two solute peaks may be separated and elute from the column at different times, the peaks may be excessively broad and may overlap with each other to such an extent that they are completely merged. In this chapter, we will investigate the effects of passage through the column on peak widths and on the resulting peak resolution.

4.1 Band Broadening

When a solute is introduced into an open-tubular column, it occupies a short but finite volume in the column as it begins to move with the carrier gas flow. When the same solute emerges from the column exit, however, it occupies a larger volume. This *band broadening* process can be severe enough to cause closely eluting solutes to largely overlap, obliterating the usefulness of the solutes' separation. The larger the separation factors between solutes, the larger is the amount of column band broadening that can be tolerated while still sucessfully resolving them. This balance between band broadening and separation is measured by the solute-to-solute *resolution* (R_s); it is one of the most important concepts to be understood about chromatographic separations.

4.1.1 Effects of Band Broadening

Band broadening can occur outside or inside the column. Sources of extra-column band broadening include the sample introduction device (the inlet), interconnections that carry the sample, and the detector at the column exit. These are considered in more detail in Parts VII and VIII of this book. For the moment, we will consider only the band broadening that occurs inside the column, and we will assume that the inlet and detector systems make no significant contributions.

Imagine a very narrow band of solute introduced into the column entrance. This band has a center of gravity about which the solute molecules are evenly distributed (see Figure 11(A) on the next page), ideally in a plug shape. As solutes traverse the column, they spend a time equal to t_R' in the stationary phase and a time equal to t_M in the mobile phase while they distribute back and forth between the two phases. Early-eluting solutes move through the column more rapidly than later eluting ones. Figure 11(B) shows the situation at time t_M, as an unretained solute elutes from the column. The initially narrow solute band has broadened somewhat; statistical averaging of the individual solute molecules' behavior produces a *Gaussian peak shape* that is typical of chromatographic separations. At the same time, other solutes are progressing through the column and are beginning to broaden.

At a time equal to t_{R1} the first retained-solute peak's center of gravity emerges from the column, as shown in Figure 11(C). The retention time is measured at the highest excursion of the

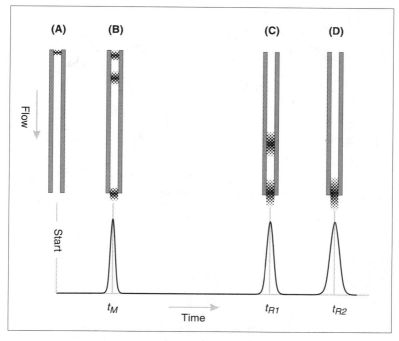

Figure 11. Peak elution from the column. Column dimensions are exaggerated for illustration purposes.

Gaussian peak (the *peak maximum*), not when the beginning or end of the peak is observed. At time t_{R2} the second peak's maximum emerges from the column (Figure 11(D) above). The result of this broadening and elution process is the chromatogram that appears across the bottom of Figure 11.

4.1.2 Measurement of Peak Width

Each peak in a chromatogram possesses a degree of broadening that can be measured in a number of ways. First, the width of the peak can be measured at a point halfway up to the peak maximum, as shown in Figure 12 on page 44. This is the *peak width at half-height*, w_h (s). The width can also be measured at the base of the peak. The peak's base is defined by the intersection of tangents to the peak's inflection points with the chromatogram's baseline. This is the *peak width at base*, w_b (s). Finally, one can also measure the *peak width at the inflection points*, w_i

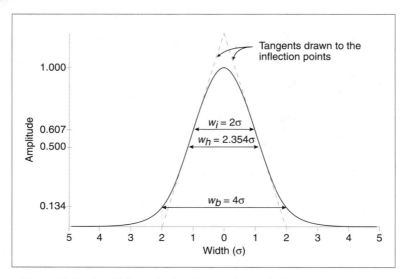

Figure 12. Peak widths and related measurements from a peak.

(s). The peak width at half-height is most often used because it is the easiest to measure directly from a chromatogram.

The peak's *dispersion* can be expressed in terms of the statistical nature of the Gaussian peak shape. The widths of a peak at half-height, at base, and at the inflection points can be related to the *standard deviation*, σ (s), of a Gaussian distribution by the following equations:

$$w_h = 2(\sqrt{2 \ln 2})\sigma = 2.354\sigma \qquad \text{eq.4.1}$$

$$w_b = 4\sigma \qquad \text{eq.4.2}$$

$$w_i = 2\sigma \qquad \text{eq.4.3}$$

The relationship of σ and the measured peak widths assumes that the peak shape is symmetical around its center of gravity. If not, then σ can be computed by determining the statistical moments of the peak shape as stored by a data handling system. In Figure 12, the x-axis is labeled in "sigma" units in order to better visualize these relationships.

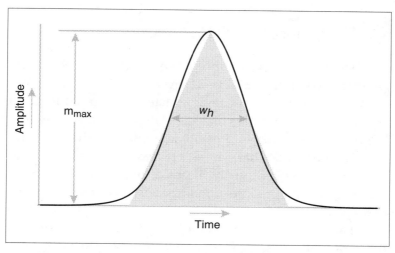

Figure 13. Approximation of Gaussian peak area from a triangle.

Peak widths can be approximated if the peak area and height are known, for example, from the printout of a chromatographic data handling system. An equilateral triangle drawn from the maximum of a Gaussian peak, through its half-height and down to the baseline has 94% of the area of the Gaussian peak itself. Figure 13 illustrates this relationship. The area of the triangle (A') is given by:

$$A' = m_{max} \times w_h \qquad \text{eq.4.4}$$

where m_{max} is the maximum peak height. Since $A' = 0.94\,A$ (where A is the true area of the Gaussian peak), we can solve eq.4.4 for w_h:

$$w_h = 0.94 A / m_{max} \qquad \text{eq.4.5}$$

Example: Calculating w_h from area and height. A peak of interest has a true area of 186 000 µV·s, and a height of 51 mV. The width at half-height is:

$$w_h = 0.94 \times (186\,000/51) \times 0.001 = 3.4 \text{ s}$$

The factor of 0.001 converts the area units of microvolt–seconds to millivolt–seconds.

4.2 The Theoretical Plate Concept

The *theoretical plate* concept is one way of expressing the ability of a column to achieve relatively narrower or wider peak shapes. The notion of theoretical plates is borrowed from the very real "plates" found in bulk distillation columns. The larger the number of plates in a distillation column, the better the boiling point "cuts" that can be obtained. A high-plate distillation column, which usually means a longer column, gives a better separation of closely boiling substances than one with fewer plates.

The same concept holds for GC: A GC column with more theoretical plates can produce narrower peaks which are better resolved from neighboring peaks than a column with fewer theoretical plates. Such a column is termed "more efficient" than the one with fewer plates. Thus, at first glance one is led to think that more plates are better. It is important to realize, however, that the theoretical plate "game" is not the only way to improve peak resolution.

The resolution of two peaks depends not only on the number of theoretical plates, but also on the separation and retention factors, α and k (see Section 6.2). Often a separation may be sufficiently improved by adjusting the column flow and temperature, or by switching to another stationary phase, instead of by increasing the plate number. This alternative usually is preferred. Achieving a higher plate count requires either longer columns which then elute peaks more slowly, or narrower columns which require higher pressure drops. The resolution "penalty" of time and/or pressure is unavoidable. It is better to adjust the stationary-phase selectivity than to resort to brute-force resolving power.

The theoretical plate concept is "theoretical" because there are no physical plates in a chromatographic column; solutes are continually distributed between the stationary and mobile phases. The notion that one plate corresponds to the average distance a solute molecule travels in one distribution step is attractive. However, it should be emphasized that the plate number and plate height concepts for chromatography represent only a descriptive way to illustrate the chromatographic process. Do not forget that it is only an approximation because an unretained peak spends all its time in the mobile phase (its molecules are not distributed into the stationary phase), yet the number of theoretical plates that it experienced in the column can be calculated the same way as for any other peak.

4.2.1 Calculating the Number of Theoretical Plates

The number of theoretical plates, N (dimensionless) can be calculated directly from a peak's standard deviation, σ, and retention time, t_R:

$$N = \left(\frac{t_R}{\sigma}\right)^2 \qquad \text{eq.4.6}$$

Narrower peaks (smaller σ values) give higher theoretical plate numbers, as expected. Based on the relationships expressed by eqs. 4.1 and 4.2, the number of theoretical plates (plate number) can also be expressed in terms of the peak width at half-height or at base:

$$N = 16\left(\frac{t_R}{w_b}\right)^2 \qquad \text{eq.4.7}$$

$$N = 5.545\left(\frac{t_R}{w_h}\right)^2 \qquad \text{eq.4.8}$$

Since the total number of theoretical plates depends upon the column length, it is useful to express column performance in terms of the distance along the column occupied by one theoretical plate. The *height equivalent to a theoretical plate, H or HETP* (mm), serves as a useful gauge of column efficiency regardless of the column length. Sometimes referred to as the *plate height*, H is calculated by dividing the column length by the total number of theoretical plates:

$$H = \frac{L}{N} \qquad \text{eq.4.9}$$

The smaller the value of H, the greater the number of theoretical plates and the more efficient the column. It will produce narrower peaks, all other factors being equal.

Another term that expresses column efficiency is the *number of theoretical plates per meter* of column length, N/L (m^{-1}). While not always included in the GC nomenclature, this term is used extensively by column manufacturers as a figure-of-merit for column testing and evaluation purposes. The number of theoretical plates per meter can be computed as the reciprocal of the *HETP*:

$$N/L = \frac{1}{H} \qquad \text{eq.4.10}$$

The number of theoretical plates per meter escalates with increasing column efficiency and is convenient for a quick comparison of column performance. Of course, as with all such comparisons, it is important to ensure that the solutes and elution conditions are the same for each column measured.

Example: Calculating the number of theoretical plates. A close-up of the last peak in Figure 2 (page 11) appears in Figure 14. This peak has a retention time of 5.3 minutes. Its width at half-height is 2.2 seconds. The number of theoretical plates is:

$$N = 5.545 \times \left(\frac{5.3 \times 60}{2.2}\right)^2 = 116\,000$$

Note that the retention time was converted to seconds by multiplying it by 60. The column in this separation is 25 meters long, so the plate height is:

$$H = (25 \times 1\,000)/116\,000 = 0.216 \text{ mm}$$

The column length is converted to mm before calculating the plate height. Finally, the number of plates per meter can be computed from H with the help of eq.4.10:

$$N/L = 1/H = 1\,000/0.216 = 4\,630 \text{ m}^{-1}$$

Here, the factor of 1 000 converts the result to plates per meter.

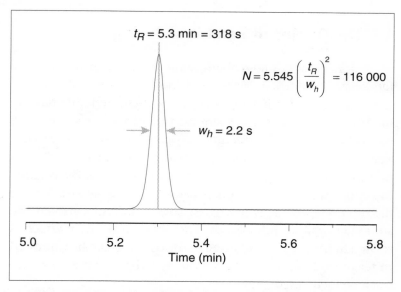

Figure 14. Measurement of theoretical plate number. Chromatogram from Figure 2, portion from 5.0 to 5.8 minutes shown.

4.2.2 Effective Plate Number

Peaks with retention factors below about 5 are in the mobile phase for 16 % or more of their passage through the column. The theoretical plate number does not account for this "dead" time; values of N for peaks with low k values are higher than they should be. The *effective plate number*, N_{eff}, subtracts the influence of the column dead time by using the adjusted retention time instead of the full retention time:

$$N_{eff} = 5.545 \left(\frac{t_R'}{w_h} \right)^2 \qquad \text{eq.4.11}$$

The effective and theoretical plate numbers converge as the retention factor increases over 5. All of the common expressions for resolution and related parameters use the theoretical plate number, and not the effective plate number. N_{eff} is not employed frequently and is appropriate only for slightly-retained peaks.

4.3 The Origins of Band Broadening

Band broadening in open-tubular columns arises from the tendency of solutes to disperse inside a column as they move from entrance to exit. This behavior is illustrated qualitatively in Figure 11 on page 43. There are two general processes that contribute to band broadening in open-tubular columns. The first arises from the tendency of solute molecules in the gas phase to diffuse away from the center of gravity of the original injection pulse. The second is caused by the kinetics of the distribution process which resist solute transfer between the stationary and mobile phases. The plate height (H), which is related to the band broadening in the column, can be expressed in terms of these two contributions and the carrier gas average linear velocity ($\bar{u}$):

$$H = \frac{B}{\bar{u}} + C \cdot \bar{u} \qquad \text{eq.4.11}$$

where B and C are the contributions from longitudinal gas-gas diffusion and resistance-to-mass-transfer, respectively. This is the *Golay equation*[8] in simplified form, and it is graphed in Figure 15. The Golay equation is sometimes referred to as the *van Deemter–Golay* equation because it is related to the van Deemter equation describing the plate height of packed columns (see Section 6.1). In Figure 15, the contributions from the B and C terms are plotted separately as functions of the average carrier gas velocity along with their sum, the theoretical plate height (H).

As the plate height in eq.4.11 gets smaller, the column operates more efficiently and generates more theoretical plates. The B term contribution decreases as the average linear velocity increases, so solutes spend less time in the mobile phase at higher velocities. At the same time, increasing the average linear velocity increases the C term contributions to the plate height.

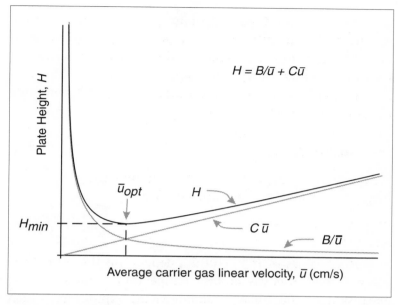

Figure 15. Plot of plate height vs. average carrier gas linear velocity for an open-tubular column.

4.3.1 The Minimum Theoretical Plate Height

The combined action of the B and C terms results in a minimum in the plot of H against $\bar{u}$ in Figure 15, at the *minimum theoretical plate height*, H_{min} (mm). The average linear velocity corresponding to H_{min} is called the optimum average carrier gas linear velocity, $\bar{u}_{opt}$ (cm/s). This is the point at which the column operates the most efficiently and produces the highest number of plates. In theory, one would set the average linear velocity at exactly $\bar{u}_{opt}$, although in practice a somewhat higher velocity is used, for two reasons. First, the slope of the H vs. $\bar{u}$ plot is steeper on the low side of $\bar{u}_{opt}$. Since the exact optimum velocity varies from one solute to another, it is better to overestimate and err on the high side. Second, when operating with a constant pressure drop across the column, the average linear velocity decreases with increasing column temperature (see Section 2.5 on page 24). For temperature-programmed elution, one must select a linear velocity which is high enough at the initial oven temperature so that the velocity at the final temperature is still at or above the optimum. Alternatively, the inlet pressure can be increased during elution to compensate for the decrease in carrier gas velocity with increasing oven temperature.

We can derive an expression for $\bar{u}_{opt}$ by differentiating eq.4.11 with respect to $\bar{u}$, setting $\frac{dH}{d\bar{u}} = 0$:

$$\bar{u}_{opt} = \sqrt{B/C} \qquad \text{eq.4.12}$$

We can derive an expression for the minimum theoretical plate height by substitution back into eq.4.11:

$$H_{min} = 2\sqrt{B \cdot C} \qquad \text{eq.4.13}$$

The exact values of H_{min} and $\bar{u}_{opt}$ depend on the solute, the column, the carrier gas, and the temperature. A more detailed expansion of the Golay equation yields expressions which permit an evaluation of column performance in terms of the best obtainable theoretical behavior.

4.3.2 The B and C Term Contributions

Contributions to the theoretical plate height from both the B and C terms are affected by the physico-chemical properties of the solute, the carrier gas, and the stationary phase. The exact nature of these dependencies in open-tubular columns is described by several equations.

The degree of contribution to theoretical plate height from longitudinal solute diffusion in the mobile phase (the B term) is controlled by the solute's *gas-gas diffusion coefficient*, D_M (cm^2/s):

$$B = 2 D_M \qquad \text{eq.4.14}$$

Solutes that diffuse more rapidly in the carrier gas will experience a larger contribution from the B term. The gas diffusion coefficient depends on the solute and mobile-phase molecular weight and size as well as the temperature and pressure of the carrier gas. For our purposes D_M is about 0.1 cm^2/s in nitrogen,

0.3 cm²/s in helium, and 0.4 cm²/s in hydrogen carrier gas. For more exact values, see ref. [24].

The C term ("resistance-to-mass-transfer" term) includes contributions both from the mobile and the stationary phases, C_M (s) and C_S (s), respectively. Their sum is equal to the overall C term contribution:

$$C = C_M + C_S \qquad \text{eq.4.15}$$

The mobile phase part of the C term depends upon the column diameter, the solute's gas-gas diffusion coefficient, and the retention factor:

$$C_M = \frac{1 + 6k + 11k^2}{96(1+k)^2} \cdot \frac{d_c^2}{D_M} \qquad \text{eq.4.16}$$

Similarly the C_S contribution depends on the stationary-phase film thickness (d_f), the solute-stationary phase diffusion coefficient, D_S (cm²/s), and the retention factor (k):

$$C_S = \frac{2k}{3(1+k)^2} \cdot \frac{d_f^2}{D_S} \qquad \text{eq.4.17}$$

The solute-stationary phase diffusion coefficient depends upon the molecular weight of the solute, as well as the temperature and density of the stationary phase and its chemical properties. It is difficult to give any single approximation for D_S because its value may range from less than 1×10^{-6} to 60×10^{-6} cm²/s or greater. Calculated D_S values for a range of solutes and stationary phases are given in ref. [24].

4.3.3 The C Term Without the Stationary-Phase Contribution

The contribution to the plate height from the C_S term is small enough to be ignored when the stationary-phase film thickness is low. Let us take a typical example:

Example: C term neglecting the stationary phase. A solute has a retention factor of $k = 8$ on a column with an inner diameter of 0.25 mm and a film thickness of 0.2 μm. The diffusion coefficients are: $D_M = 0.3$ cm^2/s and $D_S = 5 \times 10^{-6}$ cm^2/s. We can calculate the two resistance-to-mass-transfer terms using eqs. 4.16 and 4.17:

$$C_M = 2.017 \times 10^{-4} \text{ s}$$
$$C_S = 5.267 \times 10^{-6} \text{ s}$$

Their sum (eq.4.15) is 2.070×10^{-4} s. The C_M term represents 97.5 % of this value.

In such a case, eq.4.11 can be approximated by omitting the C_S term:

$$H \approx \frac{B}{u} + C_M \cdot \bar{u} \qquad \text{eq.4.18}$$

In this simplified case, we can also write $\bar{u}_{opt}$ and H_{min} by neglecting C_S. Hence, eqs. 4.12 and 4.13 can be approximated as:

$$\bar{u}_{opt} \approx \sqrt{B/C_M} \qquad \text{eq.4.19}$$

$$H_{min} \approx 2\sqrt{B \cdot C_M} \qquad \text{eq.4.20}$$

Substituting the proper expressions from eqs. 4.14 and 4.16 for B and C_M, we can express $\bar{u}_{opt}$ and H_{min} for this simplified case as:

$$\bar{u}_{opt\,(theor)} = \frac{D_M}{d_c}\sqrt{\frac{192(1+k)^2}{1+6k+11k^2}} \qquad \text{eq.4.21}$$

$$H_{min\,(theor)} = d_c\sqrt{\frac{1+6k+11k^2}{12(1+k)^2}} \qquad \text{eq.4.22}$$

We added *(theor)* to the subscript of these terms to emphasize that these are theoretical values (for the idealized case) when resistance-to-mass-transfer in the stationary phase is disregarded.

4.3.4 The C Term Including the Stationary-Phase Contribution

The C_S term is significant in the case of thick-film columns. Including it in eq.4.11 gives the following expression for the theoretical plate height:

$$H = \frac{B}{\bar{u}} + (C_M + C_S) \cdot \bar{u} \qquad \text{eq.4.23}$$

In this case, the values of $\bar{u}_{opt}$ and H_{min} would have to be calculated by considering both the C_M and C_S terms in eqs. 4.12 and 4.13:

$$\bar{u}_{opt} = \sqrt{B/(C_M + C_S)} \qquad \text{eq.4.24}$$

$$H_{min} = 2\sqrt{B \cdot (C_M + C_S)} \qquad \text{eq.4.25}$$

However, we may still use eq.4.20 (and 4.22) for a special reason; both equations establish how much our actual value differs from the ideal case (see *UTE%* in Section 4.3.6).

The extent of the stationary-phase contribution can be seen from the plots in Figure 16 on the next page (which present predicted H vs. $\bar{u}$ curves with and without the C_S contribution) along with experimental data from a column with a relatively thick, 3-µm stationary-phase film. The corresponding data is given in Table 7 on page 57. The theoretical plot without the C_S contribution (curve B) is calculated using eq.4.18. This relationship overestimates the available column efficiency by predicting too-low plate heights compared to the experimental measurements. Curve A includes the influence of the stationary film and is a much closer match to the measured values. It is calculated using eq.4.23.

> **Example: *C* term with and without the stationary-phase contribution.** Using eqs.4.14, 4.16, and 4.17, along with the constants given in the caption of Figure 16, we can calculate the values of the three terms:

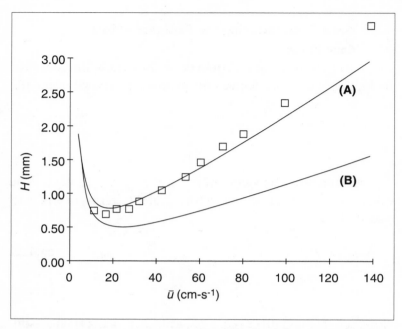

*Figure 16. Theoretical and experimental plots of plate height vs. average linear velocity for n-dodecane ($k = 8$). **(A)** Theoretical plot including the stationary phase contribution (calculated using eq.4.23). **(B)** Theoretical plot ignoring liquid film contributions (calculated using eq.4.18). **Squares**: experimental values. $D_M = 0.3$ cm²/s; $D_S = 5.0 \times 10^{-6}$ cm²/s. Column: 25–m x 0.53–mm i.d. x 3.0–µm film, 5% phenyl–95% methyl-polysiloxane on fused silica. Carrier gas: helium. Column temperature: 125 °C. See Table 7 for corresponding data.*

$$B = 0.6 \text{ cm}^2/\text{s}$$
$$C_M = 9.067 \times 10^{-4} \text{ s}$$
$$C_S = 1.186 \times 10^{-3} \text{ s}$$

The total resistance-to-mass-transfer term (eq.4.15) is 2.092×10^{-3} s. Here, the C_S term represents 57 % of the total value. It is obvious that, in the case of a thick-film column, this stationary-phase effect cannot be neglected.

We can calculate the optimum velocity and minimum plate heights for the plots in Figure 16 from the B, C_M, and C_S values given above. Curve B, which does not include the stationary-phase contribution, has $\bar{u}_{opt}$ and H_{min} values as follows:

$$\bar{u}_{opt} = \sqrt{B/C_M} = 26 \text{ cm/s}$$
$$h_{min} = 2\sqrt{B \cdot C_M} = 0.46 \text{ mm}$$

$\bar{u}$ (cm/s)	Plate height values, H (mm)		
	Experimentally measured	Theoretically calculated	
		ignoring C_S (eq.4.18)	including C_S (eq.4.23)
11.2	0.75	0.65	0.88
16.7	0.69	0.54	0.78
21.7	0.77	0.51	0.79
27.4	0.77	0.52	0.84
32.1	0.88	0.53	0.90
42.5	1.05	0.60	1.06
53.4	1.25	0.69	1.26
60.4	1.46	0.75	1.39
70.6	1.70	0.85	1.59
80.1	1.88	0.94	1.77
99.2	2.34	1.14	2.16
138.9	3.48	1.55	2.97

Table 7. *Experimentally measured and calculated theoretical plate heights with and without the stationary-phase contribution. Data from Figure 16.*

If the influence of the stationary phase is included (curve A), then C_S is included in the above equations giving:

$$\bar{u}_{opt\,(S)} = \sqrt{B/(C_M + C_S)} = 17 \text{ cm/s}$$

$$h_{min(S)} = 2\sqrt{B \cdot (C_M + C_S)} = 0.71 \text{ mm}$$

Here, the subscript *(S)* indicates that the stationary-phase contribution has been taken into account. Again, note from Table 7 that plate height and linear velocity at optimum are much closer to experimental values when the stationary-phase influence is included.

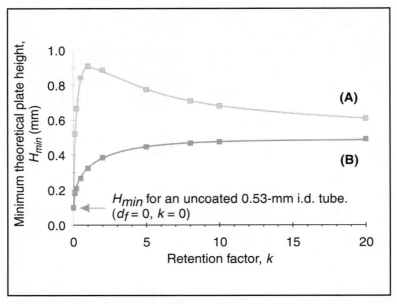

Figure 17. Influence of the retention factor on the minimum theoretical plate height. (A) Values calculated by including the stationary film contribution (using eq.4.25); (B) Values calculated by ignoring the stationary phase contribution (using eq.4.20). $D_M = 0.3$ cm^2/s; $D_S = 5.0 \times 10^{-6}$ cm^2/s. Column: 0.53-mm i.d. x 25 m x 3.0 μm film. Data for this plot appears in Table 8.

4.3.5 Influence of Retention on the Theoretical Plate Height

The retention factor also contributes to the theoretical plate height: Both the C_S and C_M terms include k in their formulas. The effects of the C_S term on the relationship of the plate height and the retention factor are illustrated in Figure 17. When the stationary-phase contribution to band broadening is not included (see plot (B) in Figure 17), the theoretical plate height increases as the retention factor increases, up to $k \approx 5$. Changing retention influences the theoretical plate height only slightly at higher k values. The situation in plot (A) of Figure 17 arises when the stationary-phase contribution is considered in the calculation of the minimum plate height. Now, H_{min} increases up to about $k = 1$, and then decreases; it approaches plot (B) at high retention. Thus, when using a thick-film column,

Retention factor, k	$HETP_{min}$ (mm)	
	$2\sqrt{B \cdot C_M}$	$2\sqrt{B \cdot (C_M + C_S)}$
0.0	0.15	0.15
0.1	0.18	0.56
0.2	0.21	0.77
0.5	0.27	1.11
1	0.32	1.24
2	0.39	1.14
5	0.45	0.87
8	0.47	0.76
10	0.47	0.71
15	0.48	0.64
20	0.49	0.62

Table 8. Data for Figure 17.

critical peaks should be placed at later retention times by choosing an appropriate column temperature.

In Figure 17, we also indicated the H_{min} value for an uncoated column tube. If no stationary phase is present then $k = 0$ and, according to eq.4.22:

$$H_{min\,(theor)} = d_c \sqrt{\frac{1}{12}} = 0.289\, d_c \qquad \text{eq.4.26}$$

This smallest possible value for H is never achieved for retained peaks. It represents a benchmark level against which all peaks can be compared.

4.3.6 Percent Utilization of Theoretical Efficiency

The measured H value (H_{meas}) can be compared to the theoretical minimum plate height ($H_{min\ (theor)}$) as a percentage:

$$U.T.E.\% = \frac{H_{min\ (theor)}}{H_{meas}} \cdot 100 \qquad \text{eq.4.27}$$

where $U.T.E.\%$ is the *Percent Utilization of Theoretical Efficiency*[34], H_{meas} is the experimentally measured plate height, and $H_{min\ (theor)}$ is the minimum plate height value calculated according to eq.4.22, ignoring the stationary-phase contribution to resistance-to-mass-transfer (the C_S term). $U.T.E.\%$ expresses the degree to which a column's *theoretical* efficiency is realized for any one peak. $U.T.E.\%$ values less than 100 % may be caused by operation at linear velocities far from optimum; by thick-film column effects as discussed above; or by a poor, uneven stationary-phase coating in the column. The $U.T.E.\%$ is also affected by the column i.d.; if the actual i.d. is significantly less than the nominal value used in calculation, the $U.T.E.\%$ may exceed 100 %. $U.T.E.\%$ values are often used by column manufacturers as a figure-of-merit for column performance.

Sometimes, the percentage value calculated acccording to eq.4.27 has been called the "coating efficiency." This is, however, an incorrect expression because differences from 100 % are not due to the "efficiency of coating" but rather to the effects described above.

> **Example: Calculating $U.T.E.\%$.** In Figure 16, the measured minimum plate height of 0.68 mm occurs at 16.5 cm/s average linear velocity. The minimum theoretical plate height calculated by neglecting the stationary-phase contribution (see page 56) is 0.45 mm. The $U.T.E.\%$ is thus:
>
> $$U.T.E.\% = \frac{0.46}{0.68} \times 100 = 67.6\ \%$$

Note that the $U.T.E.\%$ is calculated on the basis of the theoretical minimum plate height (at the optimum average linear gas velocity) compared to a measured value. This experimental value is not necessarily taken at the same gas velocity, but instead applies to an individual peak under specified conditions. The $U.T.E.\%$ value is meaningless without the conditions under which it was obtained; comparison of two $U.T.E.\%$ values *must* be carried out under identical conditions.

It is important to remember that the stationary-phase film thickness affects the solutes' retention factors, as well. As the film thickness increases, retention factors also increase as long as the column temperature is constant. The interplay of retention factor, separation factor, and plate height ultimately determine peak resolution.

4.4 Separation Quality

Resolution is critical to both qualitative and quantitative analysis, more so for the latter. For purposes of identification, two peaks must be sufficiently resolved to be distinguished from each other. Partial resolution is acceptable, but much greater resolution is required for accurate and repeatable quantitation: 95 % or higher. At lower resolutions it becomes difficult to determine which part of overlapping peaks belong to which peak, especially if the peaks are of unequal size. Selective detection (see section 8.5 on page 155) has been applied to this problem, but increased resolution remains the simplest and most practical solution.

Separation quality is expressed in a number of different ways that depend upon whether just two or many peaks are involved.

4.4.1 Resolution

The peak *resolution*, R_s (dimensionless) measures the separation quality of two adjacent peaks as the ratio of their retention time difference to their average width-at-base:

$$R_s = \frac{t_{R2} - t_{R1}}{(w_{b1} + w_{b2})/2} = \frac{2 \cdot (t_{R2} - t_{R1})}{w_{b1} + w_{b2}} \qquad \text{eq.4.28}$$

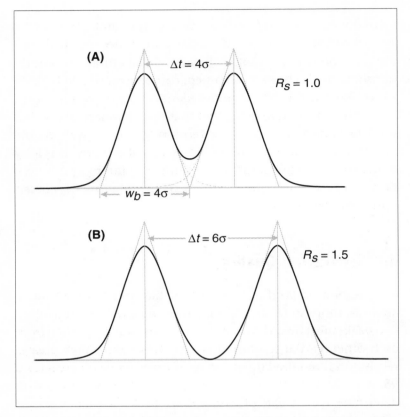

Figure 18. Different resolutions of a peak pair (A): $R_S = 1.0$. (B): $R_S = 1.5$.

where the subscripts 1 and 2 designate the first and second peaks, respectively (see Figure 18). Two peaks separated by just one width-at-base (4σ) have a resolution of 1.0, and this is the situation in Figure 18(A). If two equal-sized peaks with a resolution of 1.0 were divided at the minimum point between them and the two parts were collected as they eluted from the column, then either part would be 94–% pure, and each would contain 6 % of the other component. At resolution $R_s = 1.0$, two peaks are well-enough separated for qualitative identification. A resolution of $R_s = 1.5$ (6σ resolution) is considered "baseline" resolution and gives better than 98–% purity (Figure 18(B)). This is a good minimum resolution for accurate peak quantitation.

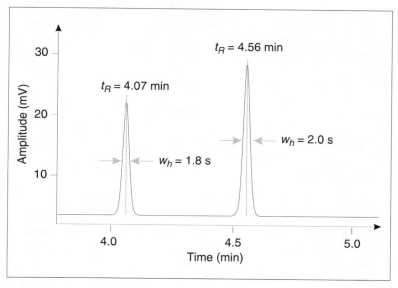

Figure 19. Chromatogram measurements used to calculate peak resolution. Section taken from Figure 2 on page 11.

Resolution calculations can be simplified in two ways. Usually, the widths of two closely spaced peaks are approximately equal so that the resolution can be expressed in terms of the width of the second peak, instead of their average:

$$R_s = \Delta t / w_{b2} \qquad \text{eq.4.29}$$

where Δt is the retention time difference between the two peaks: $\Delta t = t_{R1} - t_{R2}$.

Calculating R_s from the peak width at half-height is easier than from the base width; the width at half-height is more easily measured directly from a chromatogram. Combining eq.4.1 and eq.4.2 and substituting into eq.4.28 gives the resolution in terms of w_{h2}:

$$R_s = \frac{\Delta t}{1.699 \, w_{h2}} \qquad \text{eq. 4.30}$$

This expression for the resolution is used most often.

Example: Calculating Resolution. Figure 19 illustrates a section of the test mixture chromatogram from Figure 2 on page 11. The first peak has $w_h = 1.8$ s at $t_R = 4.07$ min, and the second has $w_h = 2.0$ s at $t_R = 4.56$ min. Calculating R_s using eq.4.28 (converting the values of w_h into w_b), we get:

$$R_S = \frac{60 \times 2 \times (4.56 - 4.07)}{(1.8 + 2.0) \times 1.699} = 9.11$$

The factor of 60 converts minutes to seconds, and 1.699 converts w_h values to w_b. If we calculate R_s using the simplified version in eq.4.30, we get:

$$R_S = \frac{60 \times (4.56 - 4.07)}{2.0 \times 1.699} = 8.65$$

Using the width of the second peak instead of the average width of the two peaks yields a slightly lower resolution value. This effect is emphasized when there is a large width difference between two peaks of interest, and in such cases, the average peak widths should be used to calculate resolution instead of the (larger) width of the second peak: eq.4.28 instead of eq.4.29.

4.4.2 Peak Number

Often, we may want to determine the total separating power of a column. While R_s measures the resolution of two adjacent peaks, calculating the maximum number of additional peaks that could be separated between two peaks conveys how complex a chromatogram could become. Measurements such as the peak number and the related separation number (see section 4.4.3) are expressions of the separation's *peak capacity* (i.e. the number of peaks which the column has the capacity to resolve[35]).

If two peaks are separated with resolution R_s, then the number of additional peaks that could fit between them with resolution R_s^* among each other is called the *peak number*, PN:

$$PN = \frac{R_s}{R_s^*} - 1 \qquad \text{eq.4.31}$$

Figure 20 illustrates the effect of changing R_s on the PN for $R_s^* = 1.5$. As the two peaks' resolution increases from 1.5 to 6, the

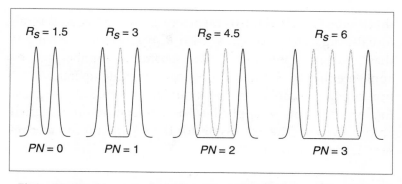

Figure 20. Peak numbers (PN) with different separation factors. $R_s^* = 1.5$.

number of additional peaks that could fit between them increases from zero to three. This calculation assumes that the extra peaks are spaced evenly between the two original peaks. Since this almost never occurs in practice, the *PN* tends to overestimate the usefulness of available space between peaks.

> **Example: Calculating the Peak Number.** For the two peaks in Figure 19, the *PN* with $R_s^* = 1.5$ is:
>
> $PN = (9.11/1.5) - 1 = 5.1$
>
> In other words, up to five additional peaks of similar width could fit with baseline resolution between the two peaks shown. This calculation used the resolution calculated from the width of the second peak. The *PN* is slightly lower if calculated from the resolution obtained using the average of the peak widths:
>
> $PN = 8.65/1.5 - 1 = 4.77$
>
> This is a more accurate estimate since, as stated above, the *PN* tends to overestimate the available space for peaks. It also accounts for the slightly larger second-peak width.

4.4.3 Separation Number (Trennzahl)

The separation number *SN* (dimensionless)* is similar to the peak number. However, it measures the number of peaks

---

* The separation number is called *Trennzahl*, *TZ*, in German.

that would fit between two members of an homologous series (usually adjacent *n*-paraffins) with $R_s^* = 1.177$, somewhat less than baseline resolution[36]. The separation number can be applied either to isothermal or programmed-temperature elution conditions, unlike the peak number which, strictly speaking, should not be applied to a temperature-programmed chromatogram. The expression for the separation number resembles that for the resolution:

$$SN = \frac{t_{R(z+1)} - t_{Rz}}{w_{hz} + w_{h(z+1)}} - 1 \qquad \text{eq.4.32}$$

where the subscripts z and $z + 1$ refer to the first and second reference peaks in the homologous series. Written in terms of the resolution of the two reference peaks, the separation number is:

$$SN = 1.177\, R_{s(z, z+1)} - 1 \qquad \text{eq.4.33}$$

✳ ✳ ✳

In this chapter, we investigated the sources of band broadening in open-tubular columns, and we developed expressions that measure the quality of a separation in terms of peak resolution. While resolution is measured on the chromatogram from retention times and peak widths, in theory it depends on the peaks' retention and separation factors, plus the column dimensions and carrier gas. In Part V, we discuss the effects of stationary-phase chemistry on retention and separation, and in Part VI, we relate the physical variables of the column to resolution and speed of analysis.

Part V

The Stationary Phase

The stationary phase is the heart of separation. Its physical and chemical characteristics determine the character of the separation. In this chapter, we will discuss various types of open-tubular columns and stationary-phase chemistries as well as their effects on separation, column preparation techniques, stationary-phase selection, and column evaluation.

5.1 The Role of the Stationary Phase

In other techniques such as liquid chromatography (LC) or supercritical fluid chromatography (SFC), the stationary and mobile phases both play major roles in the separation. Solutes interact with both phases; a change to either phase will affect the relative position of solutes as they emerge from the column. In GC, however, the mobile phase is an inert gas which acts non-preferentially towards all solutes. Changing the GC carrier gas from helium to nitrogen, for example, will not affect the relative separation of the solutes. Such a change may affect the quality of the separation as well as the optimum operating conditions, however, because of differences in carrier gas properties such as the gas-gas diffusion coefficient (see section 4.3 on page 50) and the carrier gas viscosity (see section 2.5 on page 24).

5.1.1 Solute–Stationary Phase Interaction

It is the differences in the interactions between solutes and the stationary phase which produce differences in their retention times since the inert carrier gas does not differentiate between the solutes. In general, solutes with more divergent chemical properties will experience a greater degree of separation than those which are chemically similar. There are two primary properties responsible for these interactions: the solutes' vapor pressures (at the column temperature), and the solutes' physico-chemical interactions with the stationary phase. As solutes' vapor pressures decrease, their retention times increase. For example, straight-chain paraffins with higher molecular weights have lower vapor pressures than those with lower molecular weights; higher molecular weight n-paraffins elute after lower molecular weight ones. On the other hand, physico-chemical interactions between solute and stationary phase (such as hydrogen bonding, dipolar interactions, or steric affinity) will shift solutes to later relative retention times, compared to less strongly interacting solutes.

The stationary-phase polarity is usually expressed in terms of its interaction with various standard analytes passing through the column. A classification system based on the retention index (see section 3.3 on page 35) was systematized first by Rohrschneider[30-31] and then extended by McReynolds[32]*. Recently, though, this system has not always been applied to new stationary phases. In general, nonpolar phases include non-interacting functional groups such as methyl or octyl, while polar phases contain functionalities such as cyanopropyl, hydroxyl,

---

* These stationary-phase classification systems are explained in detail in Part IX of ref. [24].

and phenyl. The nonpolar phases tend to separate solutes on the basis of their vapor pressures since there is little specific chemical interaction.

Solutes and polar stationary phases with similar chemical properties will have a stronger affinity for each other, and thus, such solutes will be more strongly retained. Carboxylic acid esters, for example, elute later on polar phases such as poly(ethylene glycol) in relation to nonpolar hydrocarbons and elute earlier on nonpolar methylsilicone phases. The selection of an appropriate stationary phase is critical for a successful separation.

Figure 21 on page 70 illustrates the effect of stationary-phase polarity on several test solutes. The chromatogram in Figure 21(A) was obtained on a nonpolar 100–% methylsiloxane phase. In Figure 21(B), a similar mixture was separated on Carbowax 20M, a poly(ethylene glycol) phase. The polar components such as 2-octanone, dimethylaniline, and naphthalene have shifted to later positions relative to the *n*-hydrocarbon peaks.

Increased stationary-phase polarity produces not only greater separation between molecules with similar vapor pressures but also different polar group compositions. For example, unsaturated fatty acid esters with the same chain length, differing only in the position of the double bond, are separated by a 50–% phenyl-, 50–% methyl-siloxane phase, but are not separated on a 100–% methyl-siloxane phase. Peaks 5 and 6 in Figure 22 on page 71 illustrate this separation for methyl octadecanoate. Figure 40 on page 146 illustrates the separation of triglycerides in butter fat according to the degree of unsaturation of the glycerine side chains on a 65%-phenyl, 35-% methyl siloxane stationary phase.

Small changes in stationary-phase polarity can help resolve two closely-eluting peaks. The 2–nitrophenol and 2,4–dimethylphenol peaks (3 and 4 in Figure 23 on page 72) are not separated on a 100–% methylsiloxane phase; the addition of 5–% phenyl substitution is sufficient to pull them apart.

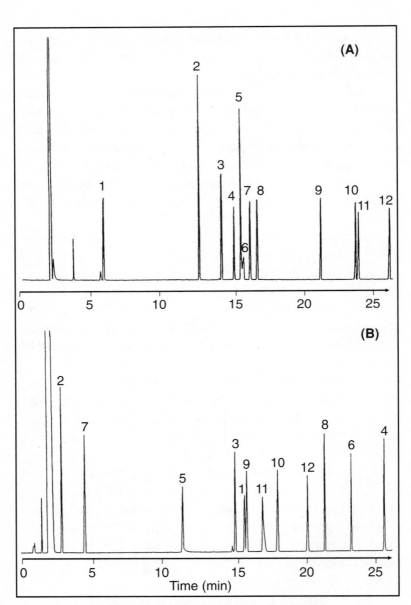

Figure 21. The effect of stationary phase polarity on relative peak positions. (A) separation on a non-polar methylsiloxane phase. (B) separation on a polar, bonded Carbowax-20M phase. **Column:** *25-m x 0.25-mm i.d. x 0.5 μm film.* **Conditions:** *oven, 40 °C, 6 °C/min to 250 °C; helium carrier, 25 cm/s.* **Peak identification:** *(1) 1-butanol; (2) decane; (3) 1-octanol; (4) 2,4-dimethylphenol; (5) nononal; (6) 2-ethylhexanoic acid; (7) undecane; (8) 2,6-dimethylaniline; (9) methyl decanoate; (10) methyl undecanoate; (11) dicyclohexylamine; (12) methyl dodecanoate*[37].

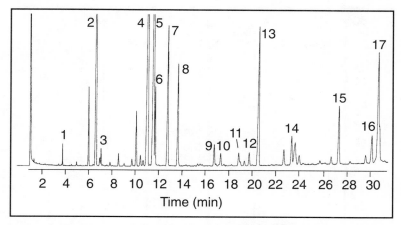

Figure 22. The separation of polyunsaturated fatty acids on a polar stationary phase. **Column:** *30-m x 0.32-mm i.d. x 0.25-μm film Stabilwax.* **Conditions:** *H_2 carrier, 40 cm/s; oven,160 °C, 2 °C/min to 250 °C for 10 min.* **Sample:** *0.1 μL split 20:1.* **Peak identification:** *(1) C14:0; (2) C16:0; (3) C16:1ω7; (4) C18:0; (5) C18:1ω9; (6) C18:1ω7; (7) C18:2ω6; (8) C18:3ω3; (9) C20:1ω11; (10) C20:1ω9; (11) C20:2ω6; (12) C20:3ω6; (13) C20:4ω6; (14) C20:5ω3; (15) C22:4ω6; (16) C22:5ω3; (17) C22:6ω3. (Stabilwax is a trademark of Restek Corp.)* [38].

5.1.2 Stationary-Phase Composition

Today, polysiloxanes (silicones) with methyl, phenyl and cyanopropyl groups are the most frequently used stationary phases in open-tubular column gas chromatography. Table 9 on page 73 lists a number of the most commonly used silicone stationary phases along with their chemical compositions. These are all polymeric silicones which are substituted with various amounts of methyl, phenyl, cyanopropyl, and vinyl groups. The overall phase polarity increases with higher levels of non-methyl substitution; the phases listed in Table 9 are in groups that follow a rough order of increasing polarity, and the numbers refer to the general chemical structure shown below:

$$-O \left[\begin{array}{c} R_1 \\ | \\ Si-O \\ | \\ R_2 \end{array} \right]_x \left[\begin{array}{c} R_3 \\ | \\ Si-O \\ | \\ R_4 \end{array} \right]_y$$

$\quad\quad\quad\quad\quad$ A $\quad\quad\quad$ B

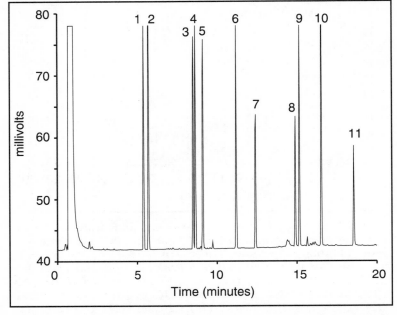

Figure 23. Analysis of phenols. **Column:** *5–% phenyl, 95–% methyl siloxane; 25–m x 0.53–mm i.d. x 1–µm film.* **Conditions:** *carrier gas, helium, 10 mL/min; oven, 2 min at 75 °C, 8 °C/min to 250 °C; packed-column inlet adapted for 0.53–mm i.d. column, 250 °C; FID, range 1, 300 °C.* **Sample:** *nominal conc. 50 ng/µL each in methanol.* **Peak identification:** *(1) phenol, (2) 2–chlorophenol, (3) 2–nitrophenol, (4) 2,4–dimethylphenol, (5) 2,4–dichlorophenol, (6) 4–chloro–3–methylphenol, (7) 2,4,6–trichlorophenol, (8) 2,4–dinitrophenol, (9) 4–nitrophenol, (10) 2–methyl–2,6–dinitrophenol, (11) pentachlorophenol.*

where A and B are two differently substituted siloxanes that are combined in proportions x and y. The distribution of the chemical functionalities along the siloxane polymer backbone has some effect on the phase polarity. For example, a 20-% phenyl, 80-% methyl phase could be synthesized from 20-% diphenylsilicone and 80-% dimethylsilicone (SPB-20) or from 40-% methylphenyl silicone and 60-% dimethylsilicone (OV-7). While the overall chemical composition (expressed as percent of the individual functional group) of these two phases is the same, the different phenyl group distributions within the polymer (in the siloxanes) gives them somewhat different chemical and chromatographic properties.

Phase Name	Phase Composition	Density 25°C/25°C	Wt-Av. Mol. Wt.	A			B		
				x	R_1	R_2	y	R_3	R_4
OV-1	Dimethyl (100%)	0.980	>10^6	100	Me	Me			
OV-101	Dimethyl (100%)	0.975	3×10^4	100	Me	Me			
OV-3	Phenylmethyl dimethyl (10% Ph, 90% Me)	0.997	2×10^4	20	Me	Ph	80	Me	Me
OV-7	Phenylmethyl dimethyl (20% Ph, 80% Me)	1.021	1×10^4	40	Me	Ph	60	Me	Me
OV-11	Phenylmethyl dimethyl (35% Ph, 65% Me)	1.057	7×10^3	70	Me	Ph	30	Me	Me
OV-17	Phenylmethyl (50% Ph)	1.092	4×10^3	100	Me	Ph			
OV-61	Diphenyl dimethyl (33% Ph)	1.090	4×10^4	33	Ph	Ph	67	Me	Me
OV-22	Phenylmethyl diphenyl (65% Ph)	1.127	8×10^3	20	Ph	Ph	70	Me	Ph
OV-25	Phenylmethyl diphenyl (75% Ph)	1.150	1×10^4	50	Ph	Ph	50	Me	Ph
OV-73	Diphenyl dimethyl (5.5% Ph)	0.991	8×10^5	5.5	Ph	Ph	94.5	Me	Me
SBP-1701	Cyanopropylphenyl dimethyl (7% Ph, 7% CP, 85% Me)			14	Ph	CP	86	Me	Me
SP-2300	Cyanopropyl phenyl (50% Ph, 50% CP)			100	Ph	CP			
SP-2310	Cyanopropylphenyl dicyanopropyl (25% Ph, 75% CP)			50	Ph	CP	50	CP	CP
SP-2330	Cyanopropylphenyl dicyanopropyl (5% Ph, 95% CP)			10	Ph	CP	90	CP	CP
SP-2340	Dicyanopropyl (100% CP)			100	CP	CP			
SE-54	Diphenyl dimethyl methylvinyl (5% Ph, 94% Me, 1% Vi)			4 Ph + 2% Ph-Vi	Ph		94	Me	Me

Table 9. Silicone stationary phase composition. Me = methyl; Ph = phenyl; CP = cyanopropyl; Vi = vinyl. Data from manufacturer's catalogs: OV = Ohio Valley; SBP and SP = Supelco; SE = General Electric.

Phase Name	Phase Composition	Typical Applications
DB-624, VOCOL, Rtx-502.2	(Cyanopropylphenyl) methylpolysiloxane	U.S.E.P.A. analyses of volatile organics in water
DB-608	Organo-siloxane	U.S.E.P.A. pesticide analysis
Carbowax 20M	Poly(ethylene glycol)	Polar separations
Molecular Sieve (PLOT)	Synthetic zeolite	Inorganic gases: H_2, O_2, N_2, CO, CO_2, etc.
Al_2O_3 (PLOT)	Inorganic salt	Light organic gases: C_1–C_5

Table 10. Specialized and non-silicone stationary phases for open-tubular columns. (DB is the trademark of J&W Scientific; VOCOL is the trademark of Supelco; Rtx is the trademark of Restek; Carbowax is a trade name of Union Carbide).

There are a number of other non-silicone phases, such as poly(ethylene glycol), that produce unique separations unobtainable with any of the silicones. Certain environmental and other analyses with specifically-mandated methodologies use separation-specific stationary phases, synthetically custom-tailored for a particular group of solutes. Some of these phases are listed in Table 10, and Figure 46 on page 160 illustrates a typical analysis on a phase designed for the United States Environmental Protection Agency (U.S.E.P.A.) analysis of volatile compounds in drinking water[39]. The exact chemical composition of such phases is not always given by the manufacturer.

Finally, there is a special group of open-tubular columns in which the stationary phase consists of an adsorbent (see section 5.2.2 on page 78 for PLOT columns). Such columns separate gases and low molecular weight organic compounds, as illustrated in Figure 24.

5.1.3 Stationary-Phase Maximum Temperatures

Modern stationary phases for open-tubular GC are usually polymeric. This was not always the case: Many phases used early in the development of GC were conveniently taken "off the shelf" from available chemical stocks and were tested for their

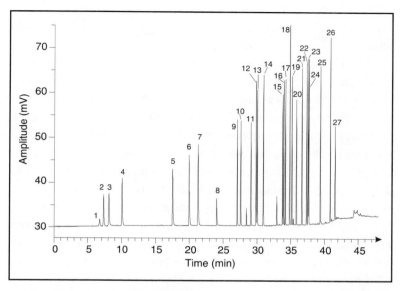

Figure 24. Separation of volatile hydrocarbons an an Al_2O_3 PLOT column. **Column:** *25-m x 0.53-mm i.d.* **Conditions:** *oven, 40 °C for 6 minutes, then 5 °C/min to 220 °C.* **Peak identification:** *(1) methane; (2) ethane; (3) ethylene; (4) propane; (5) propylene; (6) isobutane; (7) n-butane; (8) acetylene; (9) trans-2-butene; (10) 1-butene; (11) cis-2-butene; (12) cyclopentane; (13) isopentane; (14) n-pentane; (15) 2-methyl-2-butene; (16) cyclopentene; (17) trans-2-pentene; (18) 3-methyl-1-butene; (19) 1-pentene; (20) cis-2-pentene; (21) 2,2-dimethylbutane; (22) 3-methylpentane; (23) 2-methylpentane; (24) 2,3-dimethylbutane; (25) isoprene; (26) 4-methyl-1-pentene; (27)*

separation properties. Their chemistry formed the basis for many separations, but their upper useful temperatures were limited by their relatively high vapor pressures. These low molecular weight compounds, such as squalane or didecyl phthalate, would bleed off the column at elevated temperatures that today would be considered moderate. High stationary phase bleed produces high detector signal offset and noise, and causes a gradual loss of column retaining power as the stationary-phase mass in the column decreases. Columns with such phases often did not deliver the theoretical maximum plate number because the phase did not always form a smooth, consistent stationary film on the column wall (see Section 5.4 on page 82). Loss of a large percentage of the phase may also increase surface activity inside the column, leading to peak adsorption or decomposition.

The vast majority of phases used today were developed specifically for gas chromatographic use. They are prepared with high molecular weights so that they have higher practical operating temperatures than previous stationary-phase materials. Many stationary phases are also *immobilized* by chemical *bonding* to the column wall, thereby increasing their effective molecular weight and providing better masking of chemically active sites on the column wall (see section 5.4.3 on page 84). Nonetheless, all columns have rated *maximum recommended operating temperatures*, M.R.O.T. Maximum temperature values are supplied by column manufacturers and are based on testing under controlled conditions; they represent the highest continuous temperature at which the stationary phase will survive for a reasonable time. One criterion for determination of the M.R.O.T. is the level of column bleed at the detector, although there is no standard. The M.R.O.T. depends not only on the stationary-phase chemical composition but also on the degree of polymerization and the influence of residual catalytic sites on the column tube inner wall. Thus, the same type of column (phase composition, film thickness, and inner diameter) may be rated for slightly different M.R.O.T. values, depending on the manufacturer. Outside influences such as oxygen contamination in the carrier gas, chemical attack from injected samples, or sample residue build up will increase apparent column bleed levels. In general, thin-film columns will have somewhat higher M.R.O.T. values than thick-film columns.

5.2 Open-Tubular Column Types

The stationary phase is held inside the column in a number of different ways. The stationary-phase chemistry and its relationship to the column material, the desired phase ratio (β), and the film thickness (d_f) determine the choice of column

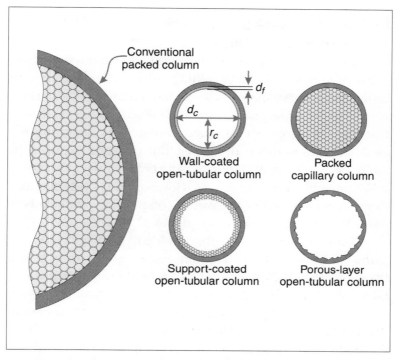

Figure 25. Types of open-tubular GC columns compared to a packed column. Drawn to scale for 0.53–mm i.d. open-tubular and 2–mm i.d. packed columns.

type. Figure 25 illustrates four common configurations and compares them to a conventional 2-mm i.d. packed column.

5.2.1 Wall-Coated Open-Tubular Columns (WCOT)

The most common column type is the *wall-coated open-tubular* (WCOT). The stationary phase is coated in a uniform film on the column inner wall. This is the simplest type of column, and it is adequate for the majority of applications. However, the film thickness of a WCOT column (and therefore the stationary-phase volume) is limited by the film stability; in the absence of polymer cross-linking and immobilization most stationary-phase films above $d_f = 0.7$ μm are not stable and will coalesce into droplets or quickly bleed off the column.

5.2.2 Support-Coated (SCOT) and Porous-Layer (PLOT) Open-Tubular Columns

The film thickness can be reduced while maintaining the stationary-phase volume by adding a finely dispersed support material to the stationary phase on the column wall[40, 41]. These are the so-called *suppport-coated open-tubular* (SCOT) columns. The effect of increasing the coated surface area is an improvement in plate number; the lower film thickness reduces the stationary-phase contribution to band broadening. The stationary film stability is also improved. However, SCOT columns may show higher adsorption of sensitive compounds, particularly in the case of nonpolar stationary phases.

Another approach to this problem involves etching a porous layer on or in the column wall before coating[42]. It is also possible to deposit a porous adsorbent layer on the inside tube wall which, in itself, acts as a stationary phase[43, 44]. Such columns are called *porous-layer open-tubular* (PLOT) columns. The net effect in PLOT columns is similar to SCOT columns: The higher inner wall surface area holds more stationary phase while maintaining a thinner film. PLOT columns are used primarily as adsorption or molecular sieve columns for open-tubular separation of low molecular weight gases and hydrocarbons such as hydrogen, helium, nitrogen, oxygen, methane, ethane, and other light gases. See Figure 24 on page 75 for an example of a PLOT column separation.

5.3 Column Materials

Many of the fundamental properties of a column are determined by the tubing material. The column inner wall must be properly deactivated so that it does not interact significantly with the intended analytes. At the same time, the wall must distribute the stationary phase uniformly; an uneven coating will cause local increases in the C_S term of the Golay equation where the stationary film is thicker, producing a less-than-optimum plate number (see Section 4.3.2 on page 52). The tubing material must be robust enough to withstand normal column operating temperatures as well as the stress of installation and routine handling. It must also form low-volume connections with other GC components to minimize extra-column band broadening. In this section, we will discuss the present state of column materials. Please refer to the Introduction (Part I) for an overview of the historical development of open-tubular columns.

5.3.1 Metal

Metal tubing is the most robust material; stainless steel, copper, and nickel tubing are flexible and strong, and they are easily cut with precise ends that form good connections. High surface activity is the principal problem with metal tubing. Stainless steel surfaces readily interact with many solutes, causing broadened or tailing polar peaks and catalyzing thermolabile peak decomposition. Nickel is somewhat more inert[45] and sometimes is used for interconnections, but avoid hydrogen carrier gas when separating aromatic or unsaturated materials in a nickel-containing GC system. Copper tubing is rarely used for open-tubular columns any more.

Metal columns do not adequately distribute stationary-phase films on untreated inner surfaces. The surface energy of the metal is too high for silicone phases, which tend to pool into droplets when coated. The addition of surfactants to the stationary phase stabilizes the film in metal columns and helps block active surface sites. Very polar phases such as *tris*-cyanoethoxypropane do somewhat better.

5.3.2 Glass

Early in the development of open-tubular columns, researchers realized that silicate glass would be a superior material in terms of its surface inertness and ability to accept stationary-phase coatings. Active surface silanol groups and metallic impurities such as boron and aluminum make glass less than ideal; a large number of surface deactivation techniques are employed to overcome these liabilities[46, 47]. It was soon discovered that acid etching of sodium glasses can produce NaCl crystals on the surface that help to better disperse the stationary phase[48–50]. The acid treatment also leaches out surface-layer metallic impurities. A number of related surface treatment techniques have been developed for glass columns[51–55]. Surface silanols are treated with alkylsilanes or are dehydrated by heating to produce Si–O–Si linkages on the surface. The resulting deactivated surface behaves well and yields minimum interactions with solutes. Deactivated glass surfaces have a low surface tension, which allows nonpolar phases to coat evenly. However, very polar phases tend not to coat well: SCOT as well as PLOT columns can overcome this problem.

Glass columns are formed into rigid thick-walled spiral shapes during production, but the column inlet and detector connections require a straightened section at either end. Column straightening requires some skill with a blowtorch so that the straightened ends maintain their internal diameter yet are straight enough for good connections. In addition, heating the column ends to the softening point after coating removes the stationary phase and exposes a fresh active surface. Once straightened, the ends often must be deactivated again. Glass is also brittle and will break easily if stressed too much or by accident. Before the widespread use of fused-silica open-tubular columns, the installation and handling of glass columns was a true art.

5.3.3 Fused Silica

During the development of glass columns it became apparent that naturally occuring quartz, with its very low metallic impurity levels, would be a superior tubing material. A very high softening point and narrow phase transition temperature range made quartz too difficult to work with; little progress occurred until the development of optical fiber drawing technology made *synthetic* fused silica the material of choice for open-tubular columns[56]. The raw material for the tubing is made from silicon dioxide that has an extremely low level of impurities, less than those found even in natural quartz. This lack of impurities gives fused silica excellent chromatographic properties but also makes it brittle once it is exposed to atmospheric moisture for a time. Therefore, fused-silica columns must be protected by an external coating, usually polyimide or aluminum, from exposure to the atmosphere and from mechanical shock. The inner surfaces are protected by deactivation pretreatments and by the stationary phase itself.

Fused-silica columns have a thin wall, on the order of 0.2–0.4 mm, which imparts flexibility*. They are produced as a single long straight piece of tubing that is coiled onto a circular cage for installation in the GC oven; column end straightening is not necessary. Fused-silica columns are easily cut off squarely by breaking the polyimide coating and lightly scoring the silica surface with a diamond, sapphire, carbide, or ceramic tool. The resulting crack propagates across the tube with slight bending pressure, producing a straight cut. The excellent mechanical properties of fused silica have virtually eliminated the need for special column installation skills and have engendered a significant expansion in open-tubular column applications.

Chemical deactivation is still required with fused silica; an untreated column exhibits high activity for neutral and basic solutes since the surface silanols act as acidic proton donor sites. Fused-silica internal surfaces are more easily deactivated than borosilicate glass column surfaces due to the lack of metallic impurities. Indeed, deactivation reagents must be very pure to avoid contaminating the inner column surface.

Fused-silica columns may be deactivated in several ways. Surface dehydration at elevated temperatures removes residual Si–OH groups by forming high energy, strained Si–O–Si bridges which readily react with silanizing reagents to form inert Si–O–

---

* It may be interesting to note that if very thin-wall capillary tubing (0.2 mm i.d.; 0.3 mm o.d.) is made of soft glass, it will also be flexible[57]. However, the initially strong tubing rapidly becomes weak and brittle. This phenomenon can only partially be overcome by application of a protective outer coating. Furthermore, such glass columns still require extensive treatment to control inner surface activity.

alkyl groups. An alternative approach involves extensive surface *hydration*, followed again by silanization, the logic being that this procedure produces better surface coverage. In either case, subsequent stationary phase cross-linking and bonding with the surface also contributes strongly to surface deactivation and shielding.

5.4 Column Coating

Once deactivated, the column inner wall is coated with the stationary phase or with prepolymers which are subsequently *cross-linked* (polymerized) *in situ*. The stationary phase must coat the column walls evenly to give the column a plate height close to the theoretical minimum. The final result depends on the stationary phase and surface qualities as well as on the coating technique.

Column heating during use may allow the stationary phase to flow and coalesce into droplets. This behavior does not depend on the coating technique but is related to the stability of the stationary film on the column surface. In general, stationary phases with surface tensions similar to the surface energy of the inner wall will be stable, while those with significantly different surface tensions tend to be unstable. Because of this, the best efforts at column coating can be spoiled upon the first heating. Chemical stabilization of the stationary-phase film by cross linking and bonding provides a good solution to the problem.

Coating the column tube with the stationary phase can be accomplished in two ways: by the dynamic or by the static coating methods.

5.4.1 Dynamic Coating

In *dynamic* coating, the column is filled with a plug of stationary-phase solution which is pushed through the column by pressurized inert gas. As the plug passes through the tubing, it deposits a thin layer of liquid. Once the plug is expelled, the remaining solvent is evaporated by purging the column with the inert gas at an elevated temperature and leaving a layer of stationary phase behind[58, 59]. It is difficult to predict the resulting film thickness exactly; it depends not only on the solution's concentration but also on the speed and viscosity of the plug as well as the surface tension of the inner wall in relation to the solution's surface tension. The *mercury drop coating technique*[60, 61] employs a short plug of mercury at the back end of the coating solution to help control the film thickness. The solution must remain stable on the column wall long enough for the solvent to evaporate, or droplets will form which degrade the eventual column performance. Dynamic coating of longer columns with higher molecular weight phases is limited by the high viscosity of the coating solution.

5.4.2 Static Coating

In *static* coating[8, 62, 63], the column is filled completely with a more dilute stationary-phase solution. One end is plugged, and a vacuum is applied to the open end. Solvent evaporates from the solution meniscus which migrates towards the plugged column end, depositing stationary phase on the column wall as it moves. Once established, the speed of motion is stabilized by the rate of evaporation at the applied vacuum and by the limit of solvent vapor flow through the column. The column must be thermostatted in order to minimize film thickness variations due to thermal expansion and contraction of the solvent plug. The resultant film thickness can be calculated from the cross-sectional column area, the solution concentration, and the stationary-phase density[5]*:

$$d_f = \frac{r_c}{2}\left[\frac{(c_L/\rho)}{100 - (c_L/\rho)}\right] \qquad \text{eq.5.1}$$

---

* For the derivation of this equation, see Appendix II in ref. [24].

where c_L is the solution concentration in wt/vol %, r_c is the inside tube radius, and ρ is the stationary-phase density. Static coating usually is preferred over dynamic coating because the film thickness can be controlled accurately. It produces a smooth initial stationary-phase film in most instances.

5.4.3 Cross Linking and Immobilization

Silicone stationary phases polymerized outside of the column can have molecular weights of over one million, although most cover the range from about 4 000 to about 40 000 (see Table 9 on page 73). However, the solubilities of the higher percent phenyl- and cyano-substituted siloxanes are not high enough to permit the preparation of adequate coating solutions. Columns cannot be prepared easily with such phases that are polymerized outside of the column. In addition, the higher molecular weight phases are solids inside the normal range of GC elution temperatures. Many columns are now prepared by *in situ* cross linking of lower molecular weight oligimers inside the column, after static coating; the final stationary phase is not formed until the cross linking is complete[65–69]. A cross-linking initiator such as a peroxide or other chemical free-radical source is included in the coating solution, or is introduced into the column as a gas after coating. Block copolymers with well-controlled compositions can be produced by including the desired proportions of prepolymers in the coating solution. Carrying out the final polymerization step inside the column reduces both the coating and the film stability problems of externally prepared phases.

Stationary phases prepared in this way may not maintain the same percent chemical composition after reaction as they had prior to coating. Reactive groups, such as vinyl and cyanopropyl, are consumed to some extent by the cross-linking polymerization process. The phase polarity may shift as a result. In addition, one must be careful to eliminate all traces of free radical initiators, such as peroxides or azides (as well as the by-products of their reactions) prior to using a newly prepared column.

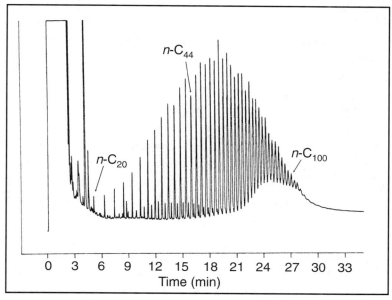

Figure 26. High-temperature chromatogram of Polywax 1000. **Column:** *8–m x 0.25–mm i.d. x 0.1–μm high-temperature methylsilicone on fused silica.* **Conditions:** *carrier gas, hydrogen, 15 psig; oven, 50 °C, 15 °C/min to 430 °C; PTV inlet, split 20:1, 50 °C initial, ballistic heating to 440 °C; detector, FID, range 1 x 8, 450 °C.* **Sample:** *0.5 μL injected, 1 mg/mL in 1:1 CS_2/n-C_{10}* [64].

The effective stationary phase molecular weight can be increased markedly by including reactive chemical functionalities in the surface deactivation step and then allowing the stationary phase to react and bond with the surface after the coating step. This chemical bonding or *immobilization* extends the operating temperature range and stabilizes the stationary-phase film on the column inner wall. It also ensures better coverage and blocking of reactive sites that could adsorb solutes during elution. Figure 26 illustrates a high-temperature chromatogram of a synthetic polyethylene wax, exhibiting peaks beyond C_{100}, with a final elution temperature of 420 °C. This kind of high-temperature analysis requires a specialized cold on-column or programmed-temperature vaporizing (PTV) injection (see section 7.6 on page 144). While stationary phases have been synthesized that are useful at temperatures up to

450 °C, few organic compounds can withstand such harsh treatment. It would seem that the true limitation of GC is no longer the stationary phase, but the sample itself.

5.5 Measuring Column Quality

Column quality is measured in several ways, but the most useful test is the intended application itself. Indeed, *good laboratory practice* (GLP) procedures require that a column's suitability for a specific separation be verified on a regular basis as part of overall system suitability testing. Typical column suitability test parameters may include plate number; separation and retention factors; resolution; Rohrschneider–McReynolds numbers; retention indices; and peak symmetry, shape, and size. These parameters must fall within minimum and maximum levels that ensure a separation of acceptable quality.

A column's performance will degrade as it ages. The rate of degradation depends on the chromatographic conditions and the care with which the column is maintained. Oven temperatures near the recommended maximum, large amounts of solvent, sample residues, poor carrier gas filtration, and lack of attention to leaking inlet septa or gas connections all can reduce the column lifetime dramatically. Periodic column verification can identify a problem before the analytical results are seriously affected.

At the column production level, the end-user's application is not always known. In practice, most manufacturers test columns with a general-purpose mixture selected according to the stationary-phase chemistry. A suite of quality control parameters are measured after production, and a column is released to stock if its performance falls within established limits.

A typical general-purpose column test mixture includes several *n*-paraffins plus various test probes with acidic, basic, aromatic, ketonic, alcoholic and other functionalities[70, 71]. We have already seen two test mixture chromatograms in this chapter: Figure 21(A) on page 70 showing a column test mixture chromatogram for a nonpolar methylsilicone phase, and Figure 21(B) showing a column test mixture chromatogram for a polar phase.

Column plate number is measured with a test solute that should elute with $k > 4$. The stationary-phase polarity and likely final application determines the probe chemistry. A n-paraffin is compatible with nonpolar or moderately polar phases, while a fatty acid methyl ester is more appropriate for a polar column. The plate number may be expressed as the total number of theoretical plates or as the number of plates per meter.

The expected plate number will be less with thicker film columns due to the larger stationary-phase contribution to band broadening. Measured column efficiencies (see section 4.3.6 on page 60 for $U.T.E.\%$) may be deceptively low if this effect is not accounted for, because the $U.T.E.\%$ does not account for stationary-phase effects on band broadening. The test peak's retention factor also must fall within set limits. At a fixed test temperature, the retention factor is a direct function of the stationary-phase film thickness; it provides a convenient verification of the coating process. A too-thin film will give lower retention factors than expected.

Polar test probes' areas or peak symmetries measure the column's adsorptive activity. On the other hand, polar peak tailing and area loss indicate specific interactions between the test probe compounds and the column tube, which may imply problems for similar polar solutes in the end user's application.

Instrumentation problems are also detected with such test mixtures since peak adsorption in the inlet system or a poor column connection may be easily mistaken for a column problem. Good testing procedures include periodic verification of the test instrument with a column of known quality.

The exact list of analytes is known for application-specific phases such as those listed in Table 10 on page 74. A resolution test with critical peak pairs from the target compound list, in addition to a general-purpose test, ensures the suitability of these special-purpose columns for their intended application.

Some manufacturers also use the retention index (see section 3.3 on page 35) of selected test probe compounds as a gauge of the column's polarity. A change of more than about two retention index units indicates a significant shift in column polarity that might affect qualitative results. Two retention index units correspond to only two percent of the distance between adjacent paraffins, or 2.4 seconds if the paraffins are two minutes apart. While this may seem like a short interval, it is in the same range as the peak width-at-base of many separations. Peaks drifting this much may be identified incorrectly.

5.6 Selecting a Stationary Phase

Stationary phase selection depends upon the intended application or applications for the column in question as well as the column's resolving power. If one column is to be used for many different applications, then a slightly polar phase and a fairly high resolution column are appropriate: for example, 5–% phenyl 95–% methlysiloxane on a 25–30 m x 0.25–mm i.d. column.

If only one separation is targeted, then the stationary phase can be selected to match the solute's requirements. The same 5–% phenyl 95–% methlysiloxane phase is a good starting point. The presence of moderate stationary-phase polarity will make little difference for nonpolar solutes such as hydrocarbons. Increased concentration of stationary phase phenyl groups will improve the separation of similar compounds that contain alcohols, ketones, esters, or aromatic groups.

A more polar phase is appropriate if the solutes differ very little in their volatilities and their polar groups. A higher phenyl content (up to 50 %) plus additional polarity in the form of vinyl or cyanopropyl groups will help the stationary phase to distinguish solutes on the basis of small structural differences.

✳✳✳

We have discussed the basics of the stationary phase and its relationship to the column in this chapter. Next, we move on to the integration of our previous discussions on retention, separation, and band broadening; we explore the manipulation of the variables of open-tubular gas chromatography.

Part VI

The Variables of Open-Tubular GC Columns

The costs of a chromatographic separation are time and pressure; the products are resolution and quantitation. That is, a given separation requires a certain time or pressure drop to achieve a desired resolution, and resolution is the prerequisite for quantitative measurement. The achievement of separation is determined by the interrelationships of the variables of the column, the carrier gas, and the temperature. Each is related to the other, and when combined with the sample's requirements for separation they produce some degree of resolution in a certain time.

Chromatograms are often over-resolved; there are large periods of baseline between most peaks. It is desirable to optimize a separation and trade off resolution for analysis time so that more analyses may be obtained in a shorter period. On the other hand, a chromatogram with insufficient resolution may require different conditions for improved resolution. Or, it may be better to choose a different stationary phase entirely (see Part V). In this chapter, we investigate the relationships of column variables, carrier gas, and temperature to resolution and speed of analysis, starting with a comparison of packed and open-tubular columns. For a more detailed discussion of the influence of column variables on chromatographic performance, see refs. [72–75] which also give further literature references.

6.1 Comparison to Packed Columns

The fundamental differences between packed and open-tubular column separations are illustrated in Figure 27. The top chromatogram in Figure 27(A) shows a separation of hydrocarbons and aromatics on a 2-mm i.d. packed column with a carrier gas flow rate of 20 mL/min. Note that peaks 7, 8, and 9 are not separated. Also, all of the aromatic peaks show a slightly tailing peak shape that is typical of solute adsorption in a packed column with a nonpolar stationary phase. Figure 27(B) shows the effects of changing to a 0.53-mm i.d. open-tubular column while keeping the same column flow rate. Now, peak 7 is beginning to be separated from 8 and 9, the overall peak shapes are very much improved, but the full separating power of the open-tubular column is not utilized. The 20-mL/min flow rate is much greater than optimum for this open-tubular column, although the high flow rate causes the analysis to take place in about the same time as with the packed column. Figure 27(C) illustrates what happens to the analysis when the flow rate is reduced to 1.4 ml/min, much closer to the optimum. The column temperature has been increased somewhat to compensate for longer retention times. All the peaks except for 8 and 9 (*m*- and *p*-xylene) are fully separated, although the analysis takes almost 24 minutes. Now, the full resolving power of the column is utilized. The generally open nature of the the 0.53-mm i.d. column enables its operation at either a high or a low flow rate and permits optimization for speed or resolution.

Measured performance data from Figure 27 is given in Table 11 on page 94. The packed-column data includes the *particle diameter*, d_p(mm) which forms a basis for comparing the

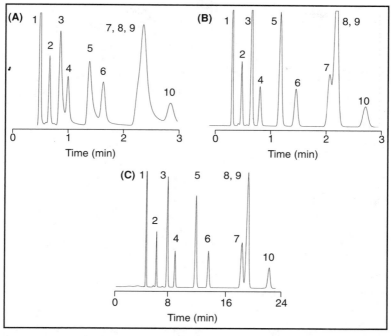

*Figure 27. Packed and open-tubular column comparison. (A) 2-m x 2-mm i.d. packed column, 8 % SE-30 on 80/100 mesh Chromosorb W-HP; helium carrier, 20 mL/min at 90 °C. (B) 25-m x 0.53-mm i.d. x 5-µm film methylsilicone fused-silica column; 20 mL/min at 90 °C. (C) same as (B), except 1.4 mL/min at 110 °C. **Peak identification:** (1) n-pentane; (2) n-hexane; (3) benzene; (4) n-heptane; (5) toluene; (6) n-octane; (7) ethylbenzene; (8) m-xylene; (9) p-xylene; (10) n-nonane.*

fundamental column resolving power to that of open-tubular columns. In packed columns[24], the C_M term is proportional to d_p^2. The van Deemter equation for the theoretical plate height on a packed column is[76]:

$$H = A + \frac{B}{\bar{u}} + C \cdot \bar{u} \qquad \text{eq.6.1}$$

where A is an additional term expressing the influence of the tortuous path of the carrier gas around the particles in the column; in a packed column, the value of A is on the order of the particle diameter. The A term is absent for open-tubular

	Packed	Wide-Bore Open-Tubular	
Column			
d_c (mm)	2.0	0.53	0.53
d_p (mm)	0.15–0.18	—	—
d_f (μm)	—	5	5
L (m)	2	25	25
T_c (°C)	90	90	110
Flow and pressure			
F_c (mL/min)	20	20	1.4
t_M (s)	20.8	14.5	220
$\bar{u}$ (cm/s)	9.6	172	11.3
Δp (psig)	30	18	≈ 1
Performance (last peak)			
t_R (s)	172	166	1 029
k	7.3	10	3.7
w_h (s)	7.9	6.2	16.9
N (plates)	2 585	3 978	20 540
N/L (plates/m)	1 298	159	822
H (mm)	0.78	2.51	1.22

Table 11. Comparative data for packed and open-tubular columns from Figure 27.

columns since there is an open, unrestricted carrier gas flow path.

The value of H_{min} for packed columns is given by an equation similar to the one giving H_{min} for open-tubular columns:

$$H_{min} = A + 2\sqrt{B \cdot C} \qquad \text{(packed columns)} \qquad \text{eq.6.2}$$

$$H_{min} = 2\sqrt{B \cdot C} \qquad \text{(open-tubular columns)} \qquad \text{eq.6.3}$$

The ratio of the minimum theoretical plate height to d_p on a packed column is expected to be larger than the ratio to d_c on an open-tubular column. This is due primarily to the A term contribution to the packed-column plate height which is absent in open-tubular columns (see the Golay equation on page 50). This is the case in Table 11; the open-tubular column (when operated close to optimum) has $H = 1.22$ mm (roughly twice the column i.d.) while the packed column has $H = 0.78$ mm, about five times the average particle diameter. The packed column is actually more efficient per unit length than the open-tubular column. This is due to the much smaller packed-column particle diameter as compared to the open-tubular column's i.d. It is not, therefore, an inherently higher efficiency per unit length (N/L) that gives the open-tubular column its greater resolving power; it is the lower pressure drop per unit length (combined with their improved inertness toward analytes) that gives open-tubular columns the potential for out performing packed columns in nearly all situations. Indeed, the only viable reasons for choosing a packed column at present are either that the methodology requires a packed column for historical reasons or that the particular separation requires a stationary phase not available on open-tubular columns.

The packed column would have to be 15.8 meters long in order to realize a theoretical plate number close to the 20 540 plates of the 25-m open-tubular column, assuming that H for the packed column remained the same and that carrier gas compression effects could be ignored. The pressure drop across such a long packed column would be over 230 psig, a prohibitively high level for standard laboratory instrumentation. Nor is it practical to prepare such a long packed column other than by coupling several 3- or 4-meter columns together, since column packing could not be introduced evenly through the entire length of such a column. In contrast, the open-tubular column pressure drop in Figure 27(C) is approximately 1 psig. A mass-flow controller regulated the column flow rate and the pressure drop could not be read accurately. If there is any problem with the open-tubular column, it is a too-low pressure drop that is easily disturbed by external influences such as injection (see section 7.2.2 on page 134).

The open-tubular column operates at low efficiency when its flow is well above optimum, in this case at 20 mL/min (Figure 27(B) on page 93). The theoretical plate height increases to 2.51 mm, about twice the value at optimum, and the total number of theoretical plates drops to about 4000. Yet this is still 1.5 times better than the packed column. The major difference at the higher flow rate is the speed of analysis: The last peak elutes in under 3 minutes on the open-tubular column and matches the packed-column time quite well. In this high flow situation, the open-tubular column separation strongly resembles the packed-column results. Efficiency has been sacrificed in favor of speed.

6.2 Relationship of Efficiency, Selectivity, and Retention

In Part IV, we introduced the theory and nomenclature of band broadening and resolution. The variables that control width and retention influence the resolution. The interrelationships of these variables gives us the ability to control and optimize resolution and time as required.

The resolution of two peaks can be expressed in terms of their efficiency (theoretical plates, N), selectivity (α), and retention factor (k). Column dimensions and linear velocity determine the fundamental efficiency (N); the stationary phase and temperature determine separation (α); and the temperature and phase ratio determine retention (k). Here, we use a relationship derived from the width-at-base and the number of theoretical plates of the second peak of the pair in question. Derivations of this and other expressions can be found in Ettre and Hinshaw[24]. A fundamental equation for peak resolution is:

$$R_s = \frac{\sqrt{N_2}}{4} \cdot \frac{\alpha - 1}{\alpha} \cdot \frac{k_2}{k_2 + 1} \qquad \text{eq.6.4}$$

where N_2 is the number of theoretical plates for the second peak, and R_s is determined concerning the peak width-at-base of the second peak (see eq.4.29 on page 63).

Here we see that resolution depends on the square root of the theoretical plate number; doubling the resolution requires four times as many theoretical plates. Concerning the retention factor, its influence is more pronounced at small values. As shown in Table 12 on the next page, 40 000 plates results only in a resolution of $R_s = 0.4$ at $k_2 = 0.2$, while the same plate number produces $R_s = 1.59$ at $k_2 = 2.0$.

The effects on resolution by increasing the separation factor (α) are more pronounced than the gains from using a longer or narrower column. The separation factor is influenced by the stationary phase and column temperature. For this reason, selecting a new stationary phase is often more effective in improving a separation than just increasing column length or decreasing column diameter in order to get more theoretical plates*.

The upper curve in Figure 28 on the next page illustrates the influence of the separation factor on resolution for a hypothetical column with 40 000 theoretical plates and $k_2 = 1.5$. Small increases in the separation factor rapidly improve resolution. In this case, when α goes from 1.05 to 1.10, R_s increases from 1.43 to 2.73. The gains in resolution from larger separation factors are much more pronounced than those from higher plate numbers.

6.3 Column Variables

Changing the column variables of length, inner diameter, and film thickness affects not only peak resolution, but also retention time and pressure drop. We can modify eq.6.4 and

---

* See section 6.6 for the influence of changing column temperatures on the separation factor.

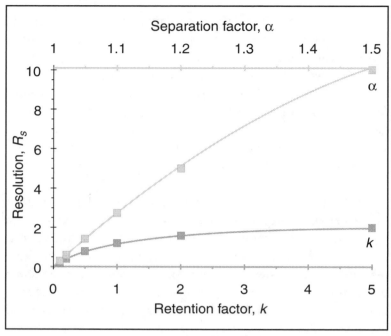

Figure 28. Influence of the separation factor and retention factor on peak resolution. N = 40 000. For the α plot, k = 1.5. For the k plot, α = 1.05. Values corresponding to the plotted points are given in Table 12.

Separation factor (α)	R_s k = 1.5	Retention factor (k)	R_s α = 1.05
1.01	0.30	0.1	0.22
1.02	0.59	0.2	0.40
1.05	1.43	0.5	0.79
1.10	2.73	1.0	1.19
1.20	5.00	2.0	1.59
1.50	10.00	5.0	1.98

Table 12. Values of R_s with different α and k values. N = 40 000. Data plotted in Figure 28.

replace the theoretical plate number with the column variables of length and theoretical plate height. Since, from eq.4.9, $N_2 = L/H_2$:

$$R_s = \frac{1}{4} \sqrt{\frac{L}{H_2}} \cdot \frac{\alpha - 1}{\alpha} \cdot \frac{k_2}{k_2 + 1} \qquad \text{eq.6.5}$$

The same square root relationship holds: Doubling the column length or halving the theoretical plate height will give only a $\sqrt{2}$ improvement in resolution.

6.3.1 Column Inner Diameter

A column's inner diameter determines the minimum theoretical plate height that the column can produce (see eq.4.22 on page 54). Other factors, such as the stationary-phase contribution to band broadening or operation above the optimum average linear velocity, will detract from this theoretical minimum. The inner diameter also determines the pressure drop required to operate the column at any specific average linear velocity (see eq.2.4 on page 20). Halving the column inner diameter will decrease the theoretical plate height by about a factor of two. This yields twice as many theoretical plates, but it is at the cost of a large pressure drop increase, since the pressure drop is inversely proportional to the square of the column diameter.

Figure 29 on the next page shows $HETP$ vs. $\bar{u}$ plots for five open-tubular columns with different internal diameters; n-undecane served as the solute and helium was the carrier gas[75]. The columns and analytical conditions are listed in Table 13. The table also lists the values of $\bar{u}$ and $HETP$ at optimum as well as the utilization of theoretical efficiency ($U.T.E.\%$) values for the columns (see Section 4.3.5 on page 58).

Decreasing the column inner diameter affects resolution, as shown for a gasoline sample in Figure 30 on page 101. The upper chromatogram was obtained on a 0.53–mm i.d. column, and the lower on a 0.10–mm i.d. column of the same length. The 5.3-fold decrease in diameter gives at least a $\sqrt{5.3} = 2.3$ times

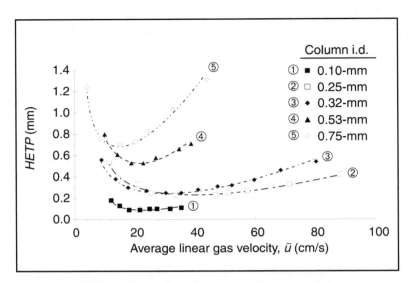

Figure 29. HETP vs. ū *plots for* n-undecane *on five open-tubular columns, operated with helium carrier gas*[75]. *For analytical conditions, see Table 13.*

i.d. (mm)	L (m)	d_f (μm)	β	T_c (°C)	k for C_{11}	$\bar{u}_{opt}$ (cm/s)	$HETP_{min}$ (mm)	UTE %
0.10	25	0.09	278	110	1.63	20.0	0.069	77.1
0.25	25	0.25	250	110	1.81	38.5	0.178	80.8
0.32	25	0.26	308	110	1.47	30.2	0.217	86.8
0.53	25	5.5	24.1	130	9.69	15.0	0.472	66.5
0.75	30	1.03	170.5	130	1.28	20.5	0.491	94.4

Table 13. Comparative efficiency data for five open-tubular columns[75].

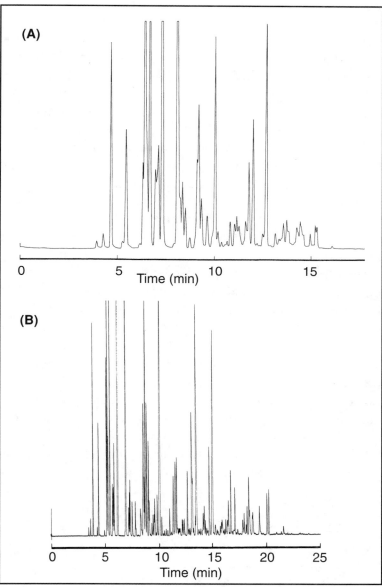

Figure 30. Effect of changing column diameter on peak resolution. (A) Column: 25–m, 0.53–mm i.d., 5–μm methylsilicone on fused silica. Carrier gas: helium at 5 psig, $\bar{u}$ = 30 cm/s. Inlet: direct, 0.05μL injected. (B) Column: 25–m, 0.10–mm i.d., 0.25–μm methylsilicone on fused silica. Carrier gas: helium at 50 psig, $\bar{u}$ = 16.7 cm/s. Inlet: split 100:1, 0.1 μL injected. **Conditions:** *oven, 30 °C, hold 2 min, 3 °C/min to 200 °C; detector, FID, 250 °C.* **Sample:** *gasoline.*

improvement in resolution. The theoretical plate height is directly proportional to column diameter if we ignore the contributions of the stationary-phase film. Additional improvement would be expected for the 0.10-mm i.d. column since, in this case, the 0.53–mm i.d. column has a thick stationary-phase film of 5 µm, while the 0.10–mm i.d. column has a thin film of only 0.25 µm. The stationary-phase contribution to the plate height will produce additional peak broadening in the 0.53–mm i.d. column, causing it to perform with much less than 100% of its theoretical potential. Meanwhile the thin-film 0.10–mm i.d. column will not experience this effect and will perform at close to 100% of its theoretical level.

Note that the 0.10–mm i.d. column requires a 50–psig pressure drop to obtain a 16.7–cm/s average linear gas velocity at the initial column temperature, while the 0.53–mm i.d. column needs only 5 psig for a 30–cm/s linear velocity. A ten-fold increase in pressure for the narrow-bore column yields only half the average linear velocity of the wide-bore column; a pressure drop in excess of 100 psig would be required for a 30–cm/s average linear gas velocity on the 0.10–mm i.d. column*. The pressure drop requirement is one limiting factor in the application of narrow-bore columns with conventional GC instrumentation, which generally has an inlet pressure drop upper limit of around 80–100 psig. This is also one reason for the use of hydrogen carrier gas with such columns, which require a lower pressure drop for the same average linear gas velocity (see Section 6.4.2).

6.3.2 Column Length

In order to maintain the average carrier gas linear velocity close to optimum, a longer column carries with it the costs of increased retention time and increased column pressure drop.

---

* This estimate is based on the relationship of pressure and linear velocity given in Section 5.5 of ref. [24].

The relationship between k, t_R, and the column length (L) is of interest, as well. Solving eq.3.4 on page 30 for t_R we obtain:

$$t_R = t_M (k + 1) \qquad \text{eq.6.6}$$

Combining eq.2.3 with eq.6.6 yields the following relationship:

$$t_R = \frac{L}{u} (k + 1) \qquad \text{eq.6.7}$$

In other words, the retention time is directly proportional to the column length for a given retention factor at the same average carrier gas velocity. Doubling the column length and increasing the inlet pressure to obtain the same average velocity will exactly double the retention times of all solutes.

The effects of changing the column length are illustrated in Figure 31 on the next page. A 50–m x 0.25–mm i.d. column was repeatedly cut in half after running a test chromatogram, while keeping the average linear velocity nearly constant at about 32 cm/s by decreasing the pressure drop as required. As the column length decreases from 50 to 2.5 meters, the retention time of the last peak drops in proportion. The peak data, including the resolution of two of the peaks, is given in Table 14 on page 105.

These chromatograms are a good example of the tradeoffs between column length, retention time, and resolution. We can ask the question, "What is the shortest column that will still baseline resolve the two closest peaks (2-octanone and n-decane)?" Starting with the 50–m column, the chromatogram is over-resolved: There are large empty areas between peaks. As we shorten the column, the closest peaks begin to overlap as we switch from the 5–m to the 2.5–m column length. Thus, it is possible to resolve this test mixture on a 5–m column in just over one minute, compared to the over-resolved separation on the 50–m column that takes more than 15 minutes. In theory, the resolution gained by doubling the column length is in proportion to $\sqrt{2}$ (see eq.6.5), but eluting the peaks requires twice as much time.

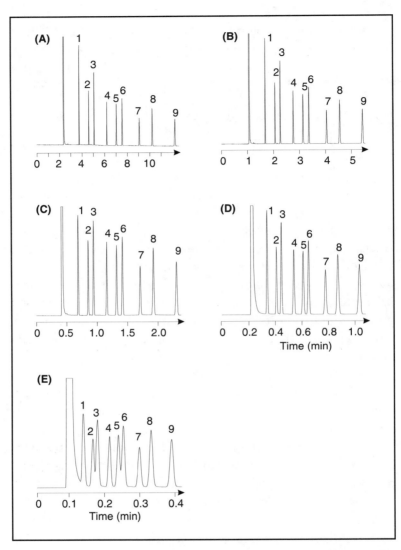

Figure 31. The effect of column length on peak resolution. **Columns:** *(A) 50–m, (B) 25–m, (C) 12–m, (D) 5–m, and (E) 2.5–m, 0.25–mm i.d., 0.25–μm methylsilicone on fused silica.* **Conditions:** *carrier gas, helium; oven, 100 °C; inlet, split, 50 mL/min, 250 °C; detector; FID, range 1x4, 250 °C, signal normalized to peak 1, except (E).* **Peak identification:** *(1) n-nonane, (2) 2-octanone, (3) n-decane, (4) 1-octanol, (5) 2,6-dimethylphenol, (6) n-undecane, (7) 2,4-dimethylaniline, (8) naphthalene, (9) n-dodecane. Data from these chromatograms is given in Table 14*[77].

L (m)	t_M (min)	$\bar{u}$ (cm/s)	t_R, n-C_{12} (min)	R_s 2-octanone/n-C_{10}
50.4	2.83	29.7	15.33	11.24
24.9	1.38	30.1	6.88	7.65
12.5	0.58	35.8	2.89	3.71
4.9	0.26	32.0	1.29	2.42
2.5	0.12	35.2	0.50	0.86

Table 14. Data for various column lengths from Figure 31.

6.3.3 Stationary-Phase Film Thickness

The film thickness affects both retention and sample capacity. Selection of an appropriate film thickness is determined primarily by the overall sample volatility. The phase ratio, β, defines the relationship between the retention factor, k, and the distribution constant, K:

$$K = \beta \cdot k \qquad \text{eq.6.8}$$

According to eq.6.8, smaller β values increase k; as we have seen in eq.6.7, an increase in k increases the retention time. Conversely, an increase in the phase ratio will decrease both k and t_R. However, this situation is made more complicated by the influence of k on peak resolution (see eq.6.5 on page 99).

The phase ratio of a column is determined by the inner tube diameter (d_c) and the stationary-phase film thickness (d_f, see eq.3.6 on page 31):

$$\beta = \frac{d_c}{4\,d_f} \qquad \text{eq.6.9}$$

The influence of the film thickness is the more significant of these two parameters because it is the easiest to control (film thicknesses range from 0.1 to 5 µm in practice). Some general guidelines for film thickness selection follow.

From eqs. 6.8 and 6.9, an increase in film thickness will decrease the phase ratio and will increase the retention factor. Thus, *thick-film columns* with relatively low phase ratios are most often used for the separation of volatile compounds that have relatively low retentions. The higher retention factors caused by a thicker film improve peak resolution, as long as the longer retention times are not detrimental. Volatiles such as pentane, hexane, and heptane are easily separated on thick-film columns without subambient oven temperatures; thin-film columns require reduced oven temperatures in order to increase retention factors and separate volatile compounds.

The increased retention of later peaks on thick-film columns may be offset by temperature programming the column oven. However, there are limitations because strongly retained peaks may require oven temperatures close to or above the stationary phase maximum recommended temperature, for elution within a reasonable time period. Therefore, thick-film columns are generally not suitable for the analysis of higher boiling solutes or samples that encompass a wide-boiling-range mixture including high-boiling compounds.

Instead, such mixtures generally are analyzed on *thin-film columns*. Here, the higher phase ratio reduces the retention factor and produces shorter retention times or lower elution temperatures. In the case of polynuclear aromatic hydrocarbons (PAH), for example, a thin-film column with a high phase ratio helps to increase the speed of analysis while limiting the final elution temperature and prolonging column lifetime. Reduced elution temperatures are also important in the case of biologically significant compounds which may decompose under prolonged exposure to higher temperatures.

We have discussed the practical influence of stationary-phase film thickness. Looking at it from a purely theoretical point of view, however, we cannot forget that the C_S term, expressing resistance-to-mass-transfer in the stationary phase, is related to the square of the film thickness (see eq. 4.17 on page 53). An increase in film thickness will increase the C_S term and reduce column efficiency. Therefore, the selection of the proper film thickness will always represent a compromise between theoretical and practical considerations.

6.3.4 Influence of Column Variables on Sample Capacity

In general, the sample capacity refers to the amount of a solute that can be tolerated in the column before efficiency is significantly reduced and peak shapes become distorted. There are no general guidelines as to how much efficiency loss and peak distortion can be tolerated, due mainly to the great variety of samples that can be analyzed. We can state, however, that chromatographic deterioration may be tolerated in practice as long as the two closest peaks maintain a minimum level of resolution. In this section we investigate the general influence of various column parameters on sample capacity.

Based on their fundamental paper published in 1956[76], van Deemter *et. al.* expressed the limiting amount of solvent vapor (V_{max}) which may enter the column as:

$$V_{max} = a_k \, v_{eff} \sqrt{N} \qquad \text{eq.6.10}$$

where N is the number of theoretical plates and v_{eff} is the so-called effective volume of a plate; a_k is an undefined correction factor. Further substitutions lead to the following relationship:

$$V_{max} = a_k \frac{V_G (1+k)}{\sqrt{N}} \qquad \text{eq.6.11}$$

where V_G is the gas-phase volume in the column which can be calculated from the column dimensions:

$$V_G = \left[\frac{d_c}{2} - d_f\right]^2 \pi L \qquad \text{eq.6.12}$$

If $d_c \gg d_f$, which is the case for most open-tubular columns, then:

$$V_G \approx \frac{d_c^2 \cdot \pi \cdot L}{4} \qquad \text{eq.6.13}$$

These relationships cannot be used to calculate the absolute value of column sample capacity because of the uncertainty of the factor a_k in eqs.6.10 and 6.11. The subjective nature of the sample capacity also makes absolute evaluation difficult. However, the sample capacity can be applied, in a relative sense, to changes that occur when the column parameters are modified.

In the first case, we will assume that the tube diameter (d_c) and the column length (L) remain constant while the stationary-phase film thickness (d_f) increases. In this case, the phase ratio (β) decreases and the retention factor (k) increases while the column gas volume (V_G) remains nearly unchanged. From eq.6.11, it follows that the sample capacity will be influenced primarily by relative changes in the factor ($k + 1$). If k is small, for example 0.1 (for early eluting peaks on a thin-film column), then this influence will not be very significant. Doubling the film thickness and doubling k only changes ($k + 1$) from 1.1 to 1.2. If k is larger, for example 10, then doubling the film thickness will increase ($k + 1$) from 11 to 21. Thus, increasing the film thickness alone will enhance the sample capacity for later eluting peaks.

In the second case, we will assume that the film thickness and column length are held constant while the column diameter is increased. The gas volume increases by the square of the column diameter increase (see eq.6.13) while the phase ratio, and thus k, decreases linearly in proportion to the increasing diameter (see eq.6.9). Thus, the overall influence of increasing the column diameter is also to increase the sample capacity, but more for early eluting peaks than for late eluting peaks.

From these considerations, the best approach towards increasing the sample capacity of an open-tubular column is to increase *both* the tube diameter and the stationary-phase film thickness, taking care that the phase ratio decreases somewhat overall. This is best accomplished by a higher relative increase in film thickness over the tube diameter.

Eq.6.11 includes the square root of the theoretical plate number (N) in the denominator, and its influence also must be

Column i.d. (d_c) mm	Film thickness (d_f) mm	Phase ratio (β)	Retention factor (k)	Plate number (N)	Retention time (t_R) s	Relative sample capacity
0.10	0.10	250	0.50	480 000	225.0	0.10
0.10	0.25	100	1.26	368 550	338.7	0.17
0.25	0.25	250	0.50	192 000	225.0	1.00
0.32	1.00	80	1.58	102 080	386.4	3.83
0.53	1.00	132	0.94	70 420	291.5	9.58
0.53	5.00	26	4.79	43 940	868.4	35.03

Table 15. *Comparative theoretical sample capacity data for six open-tubular columns with varying tube diameter and film thickness. Column length is 30 m in all cases. Plate number refers to a peak with the given retention factor. Resolution is calculated considering the peak width and retention factor of the second peak of a pair with $\alpha = 1.05$. The retention factors refer to the same solute under the influence of changing phase ratio. The average linear gas velocity ($\bar{u}$) is 20 cm/s in all cases.*

considered. Increasing the film thickness or the tube diameter will reduce the number of theoretical plates and give a corresponding increase in sample capacity. The square root relationship makes this effect much smaller than the influence of the other factors.

Table 15 gives theoretical relative sample capacities for six open-tubular columns. In this calculation we compare the value of this fraction from eq.6.11;

$$\frac{V_G (1 + k)}{\sqrt{N}}$$

for the listed columns, considering its value for a 0.25–mm i.d. column with a 0.25–µm film thickness as 1.00. From the table, we see that the same length of 0.10–mm i.d. column with a film thickness of 0.10 µm would have one-tenth the sample capacity of the reference column. The sample capacity of a corresponding column with 0.53–mm i.d. and a 5–µm film thickness would be 35 times the sample capacity of the reference column.

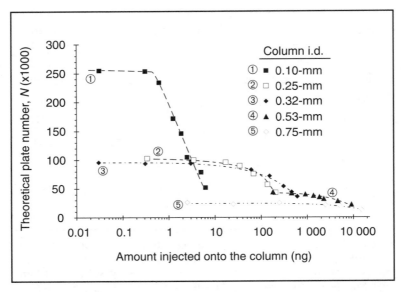

Figure 32. Number of theoretical plates calculated for n-undecane as a function of the injected sample amount. The column i.d. is indicated in the figure. All data refers to 25–m column lengths. See Table 13 on page 100 for the analytical conditions[75].

It is also interesting to consider sample capacity data measured from real columns. We will define the sample capacity as the amount of solute that produces a 20-% reduction in the number of theoretical plates on any individual column, as we increase the injected amount. Here, we use the same columns and conditions already discussed in Figure 29 and Table 13 on page 100. The results are shown graphically in Figure 32, which plots measured theoretical plate numbers as injected sample amounts increase. The sample capacity values from the plots are listed in Table 16. These results indicate that — depending on the column parameters — the sample capacity of open-tubular columns for a single solute varies from about 0.1 ng to 2–3 µg.

L (m)	d_c (mm)	d_f (µm)	β	T_c (°C)	Sample capacity (ng)
25	0.10	0.09	278	110	0.9
25	0.25	0.25	250	110	52.9
25	0.32	0.26	308	110	114.5
25	0.53	5.50	24.1	130	2 500
30	0.75	1.03	130	130	2 000

Table 16. Measured sample capacity values for five open-tubular columns coated with methylsilicone stationary phase. Solute: n-undecane. Sample capacity is defined as the solute amount corresponding to a 20-% reduction in the theoretical plate number[75].

6.4 Carrier Gas Variables

The variables of the carrier gas — temperature, velocity, flow, and gas type — can have a profound influence on chromatographic performance. In this section, we will discuss these variables and explore their effects.

6.4.1 Carrier Gas Velocity

Carrier gas velocity influences both the efficiency and the speed of analysis. Column efficiency degrades as the average linear gas velocity ($\bar{u}$) departs from its optimum value (as illustrated for several columns in Figure 29 on page 100). According to theory, and as observed in practice, the descending slope of a plot of *HETP* vs. $\bar{u}$ at velocities below optimum is steeper than the ascending portion above the optimum velocity (see Figure 16 on page 56). Therefore, carrier gas velocities somewhat above optimum are preferred for two reasons. First, the optimum velocity is somewhat different for each solute, due to differences in the solute's retention factors and diffusion coefficients. Second, column temperature changes affect the linear gas velocity. It is better to operate in a region where small changes in gas velocity have little effect on efficiency and where the gas velocity is always above the optimum value for each solute in the separation.

The speed of analysis (t_R of the last peak) is inversely proportional to the average linear gas velocity when the column is operated at a constant temperature. We have already mentioned this relationship in the discussion of the effects of the column length (see page 103):

$$t_R = \frac{L}{\bar{u}}(k+1) \qquad \text{eq.6.7}$$

Here, if we hold the column length and the retention factor constant, doubling the average linear gas velocity will cut the analysis time in half. On the other hand, the increased speed may carry with it the penalty of reduced efficiency as the average gas velocity moves well above the optimum velocity. This is an unavoidable tradeoff when taking this approach to achieving faster analysis times. However, the optimum gas velocity depends not only on the solutes, but also on the type of carrier gas used for the separation. Therefore, it is important to understand the influence of the carrier gas; by selecting a carrier gas with a higher optimum velocity, it is possible to decrease the analysis time with little or no sacrifice of column efficiency.

6.4.2 Carrier Gas Type

Only three carrier gases are used in nearly all GC analyses: nitrogen, helium, or hydrogen. There are other choices, but these three constitute the vast majority. The type of carrier gas influences column performance in two principal areas. The first is related to the value of the optimum average linear gas velocity ($\bar{u}_{opt}$). Its value is proportional to the gas–gas diffusion coefficient (D_M) of a solute in the carrier gas; higher D_M values give higher optimum linear velocities (see eq.4.21 on page 54). The value of D_M for a given solute is lowest in nitrogen and highest in hydrogen.

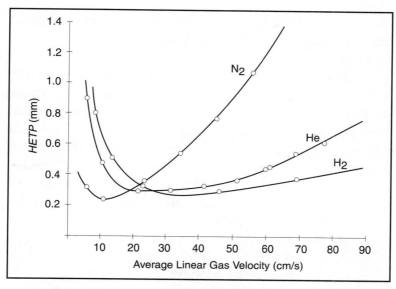

Figure 33. HETP vs. $\bar{u}$ plots for n-heptadecane at 175 °C, for three carrier gases. Column: 25-m x 0.25-mm i.d., 0.4 μm OV-101 methylsilicone on fused silica [78].

Figure 33 illustrates *HETP* vs. $\bar{u}$ plots for three carrier gases. The measured values of $\bar{u}_{opt}$ are 12 cm/s for nitrogen, 22 cm/s for helium, and 36 cm/s for hydrogen. We can conclude from these measurements that the analysis time at optimum gas velocity using nitrogen will be 83 % longer than with helium. The analysis time with hydrogen carrier gas will be only 61 % of the time required with helium. Thus, the carrrier gas with the highest gas-gas diffusion coefficient also produces the fastest analysis times.

The second major influence of the carrier gas type is in the slope of the ascending part of the *HETP* vs. $\bar{u}$ plot at velocities above $\bar{u}_{opt}$. Nitrogen produces the highest slope in this area, followed by helium and then hydrogen. Operation above the optimum average linear gas velocity with nitrogen carrier will significantly increase the *HETP*. On the other hand, operation in this region with hydrogen carrier only has a small effect on efficiency since the slope of the *HETP* plot is small; velocities well above optimum can be selected with little or no significant losses in column efficiency. Column efficiency losses with helium carrier gas above the optimum are nearly as low as for hydrogen carrier.

Carrier gas selection has two additional consequences. The first is related to the value of $HETP_{min}$. As can be seen in Figure 33, the value of $HETP_{min}$ is about the same for helium and hydrogen carrier gas — 0.29 vs. 0.27 mm — but it is smaller for nitrogen carrier gas — 0.23 mm. Column efficiency is higher with nitrogen carrier gas at the optimum velocity than with either hydrogen or helium, by about 20 %. This slight advantage with nitrogen can be compensated with helium or hydrogen by using a somewhat longer column, which will produce the additional theoretical plates, and by operating above the optimum gas velocity in order to recover the speed of analysis that would otherwise be lost on a longer column.

The second additional consideration for carrier gas selection is the effect on inlet pressure. In Section 2.2 on page 18, we discussed the relationship between pressure drop (Δp), average carrier gas velocity ($\bar{u}$), and carrier gas viscosity (η). As shown in eq.2.4 on page 20, Δp is proportional to the product $\bar{u} \cdot \eta$. On the other hand, the viscosity of hydrogen differs significantly from both helium and nitrogen, as shown in Figure 5 on page 25. Thus the pressure drops associated with hydrogen carrier gas are significantly lower than those for helium or nitrogen. This is advantageous for operation at higher carrier gas velocities or with smaller-i.d. columns.

Example: Calculating pressure drops with different carrier gases. We consider a 30-m x 0.25-mm i.d. open-tubular column operated at 100 °C with three carrier gases at average linear gas velocities of 12 cm/s (nitrogen), 22 cm/s (helium), and 36 cm/s (hydrogen). The viscosities of the three gases at 100 °C are (see Section 2.5):

nitrogen: 2.057×10^{-5} Pa·s
helium: 2.266×10^{-5} Pa·s
hydrogen: 1.023×10^{-5} Pa·s

The pressure drop is calculated using eq.2.4:

$$\Delta p \approx \frac{32 \cdot L \cdot \eta \cdot \bar{u}}{d_c^2}$$

Using the proper dimensions (L and d_c in cm, $\bar{u}$ in cm/s, and η in Pa·s), Δp will be in units of Pascals. Multiply by 1.4504×10^{-4} to convert to psig. The following values are obtained for the pressure drop:

nitrogen: $\Delta p = 5.50$ psig
helium: $\Delta p = 11.10$ psig
hydrogen: $\Delta p = 8.20$ psig

6.5 Speed of Analysis

The speed of analysis refers to the time required to separate two peaks with resolution R_S and separation factor α. A faster speed of analysis is desirable when more sample throughput is required. We have seen from eq.6.7 that:

$$t_R = \frac{L}{\bar{u}}(k+1) \qquad \text{eq.6.7}$$

Thus, the easiest way to decrease the analysis time is to increase the carrier gas velocity. However, as we saw in section 6.4, column efficiency is reduced at higher gas velocities. We also can change other column variables to compensate for such efficiency losses if they cause peak resolution to drop below acceptable levels. One possibility is to use another carrier gas; changing from helium to hydrogen permits operation at linear velocities about 50 % higher with no efficiency sacrifice.

We can also reduce the column inner diameter in order to realize a higher optimum gas velocity. From eq.4.21 on page 54, the theoretical optimum velocity is inversely proportional to the column i.d.:

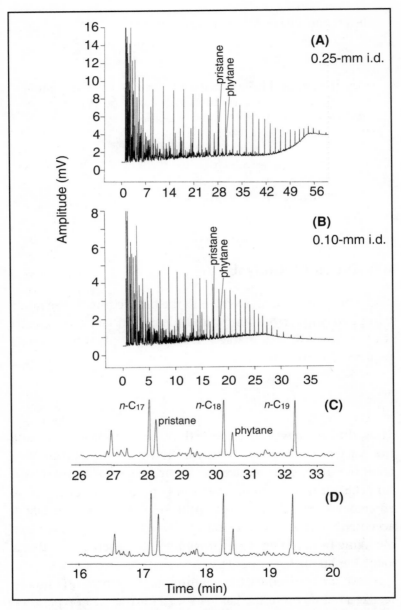

*Figure 34. Influence of column diameter on speed of analysis. (A) 25-m x 0.25-mm i.d., 0.25-μm film methylsilicone on fused silica. Helium, 25 psig, $\bar{u}$ = 33 cm/s; (B) 15-m x 0.10-mm i.d., 0.25-μm film. 66 psig, $\bar{u}$ = 33 cm/s; (C) portion of (A); (D) portion of (B). **Sample**: 0.1 μL crude oil. **Conditions**: inlet, PTV split, 35 °C to 400 °C, 50 mL/min split vent flow; oven, 35 °C, 5 °C/min to 300 °C.*

$$\bar{u}_{opt(theor)} = \frac{D_M}{d_c} \sqrt{\frac{192(1+k)^2}{1+6k+11k^2}} \qquad \text{eq.4.21}$$

This relationship assumes that the stationary-phase influence on the plate height is negligible and that the column phase ratio, and hence the retention factor, remain constant. If we reduce the column i.d. from 0.53-mm to 0.10-mm, then we should achieve a 5.3-times faster optimum linear velocity. The analysis time would be reduced by about 80 % if the lengths of the two columns were the same.

In practice, there are other factors that must be considered. A narrower bore tube will have an inherently better efficiency per unit length; therefore we can reduce the column length as well as the i.d. and maintain the same total number of theoretical plates. Length reduction is advantageous because it not only further reduces the analysis time, but it lowers the inlet pressure required for operation of narrower i.d. columns at higher linear gas velocities. Here, hydrogen carrier gas is doubly advantageous, affording lower pressure drops that compensate for higher pressures required by faster optimum linear gas velocities and narrower column inner diameters.

The influence of the column diameter on the speed of analysis is illustrated in Figure 34. A crude oil sample was injected onto a 25-m x 0.25-mm i.d. column and onto a 15-m x 0.10-mm i.d. column with the same film thickness. Both columns were operated at the same linear velocity under the same temperature conditions. Note that the 0.10-mm i.d. column required a pressure drop of 66 psig, while the 0.25-mm i.d. column needed only 25 psig for the same average linear gas velocity. Figure 34(A) shows the chromatogram on the 0.25-mm i.d. column; the total analysis time for the phytane peak was 30.5 minutes. The chromatogram for the 0.10-mm i.d. column in Figure 34(B) shows the phytane peak eluting in 18.3 minutes; the analysis time on the 0.10-mm i.d. column was 60 % faster. The overall column efficiencies were about the same for the two columns. Figure 34(C) and (D) show a closeup of the portions of each chromatogram containing the C_{17} through C_{19} peaks.

The resolution on the 0.10-mm i.d. column is just slightly better than on the 0.25-mm i.d. column.

We can express the general influence of the column efficiency (H), average linear gas velocity ($\bar{u}$), retention factor (k), separation factor (α), and resolution (R_s) on the speed of analysis by combining eqs. 6.5 and 6.6:

$$t_{R2} = 16 R_s^2 \left(\frac{\alpha}{\alpha - 1} \right)^2 \cdot \frac{(1 + k_2)^3}{k_2^2} \cdot \frac{H_2}{\bar{u}} \qquad \text{eq.6.14}$$

In this equation, the influence of the last two terms is particularly important[79]. The term $(1 + k_2)^3 / k_2^2$ has a minimum at $k_2 = 2$. This means that, whenever possible, one should try to adjust the column parameters and analytical conditions so that peaks representing the most critical separation have retention factors close to this value.

We have to reconsider the Golay equation (eq.4.11 on page 50) in order to realize the significance of the last term:

$$H = \frac{B}{\bar{u}} + C \cdot \bar{u} \qquad \text{eq.4.11}$$

The influence of the $B/\bar{u}$ term is negligible at velocities well above optimum, as shown in Figure 15 on page 51, and eq.4.11 can be approximated by a linear equation:

$$H \approx C \cdot \bar{u} \qquad \text{eq.6.15}$$

so that

$$C \approx \frac{H_2}{\bar{u}} \qquad \text{eq.6.16}$$

In other words, the last term in eq.6.14 is equal to the slope of the Golay plot at higher linear gas velocities. This emphasizes again the importance of carrier gas selection with respect to the speed of analysis; the smallest slope is obtained with hydrogen carrier gas, as discussed in Section 6.4.2. Thus, hydrogen permits the highest relative speed of analysis.

6.6 The Effect of Column Temperature

In GC, the column temperature either is held constant (*isothermal* operation), or is increased while solutes are traveling down the column (*programmed-temperature* operation).

6.6.1 Isothermal Elution

Isothermal operation is generally used unless the range of solute retention times is greater than about 30 minutes. Solute retention times are strongly affected by the column temperature; higher isothermal temperatures give shorter retention times. Normal column operating temperatures cover a wide range in open-tubular GC, from below room temperature up to 350 °C or even higher in certain applications, allowing a very wide range of solutes to be separated. Unlike the linear relationships of retention with the phase ratio, carrier gas velocity or column length, there is an *exponential* relationship between the isothermal column temperature, T_c, and solute retention. In general, the retention time is cut in half for every 12 to 15 °C increase in the column temperature. The relationship between the distribution constant (K) and the isothermal column temperature can be expressed as:

$$\log K = \frac{A}{T_c} + B \qquad \text{eq.6.17}$$

where A and B are arbitrary constants, and the column temperature (T_c) is expressed in degrees Kelvin. As T_c increases, K decreases and therefore, the retention factor (k) as well as the retention time will also decrease.

The effect of increasing the isothermal column temperature is similar to the effect of decreasing the film thickness: peaks elute at earlier retention times. There is one critical difference, however; the values of the constants A and B in eq.6.17 are different for each solute. This means that the *relative* position of some peaks in the chromatogram will change as the temperature changes. At one temperature two peaks may be separated, but at another they may elute at the same time, or their elution order may even reverse if the temperature difference is large enough. This effect is illustrated in Figure 35, which shows the separation of four hydrocarbons on a nonpolar column at three different temperatures. The methylcyclohexane peak moves relative to the other peaks as the temperature increases. It is the first peak at 45 °C, the second peak at 75 °C, and the third peak at 95 °C. This effect also causes it to coelute with other peaks at certain temperatures. It will have the same retention time as peak 2 at a temperature between 45 °C and 75 °C and the same retention time as peak 3 between 75 °C and 95 °C. At or near these critical temperatures, the coeluting peaks will appear as one peak, or as a broad but unresolved peak. For this reason, when there is any possibility of peak coelution, the separation temperature conditions should be varied in order to reveal the "hidden" peaks. An even better approach involves separating the solute mixture in question on a column with a different polarity stationary phase. The chances of coelution on both columns are slight. A mass spectrometric detector also helps to determine the purity of a suspect peak. A mixed mass spectrum will indicate the presence of an impurity.

When peaks' relative positions in a chromatogram change, so does their separation factor (α). The relationship between temperature and separation factor can be expressed by an equation similar to eq.6.17. Over a relatively short temperature range, we can write that[80]:

$$\alpha = aT + b \qquad \text{eq.6.18}$$

where a and b are different constants. If a is positive then the separation factor will increase with increasing temperature, while if it is negative then the separation factor will decrease. Some examples of this relationship are shown in Figure 36 on page 122.

In isothermal elution, one should try to place the most critical peak pair at a retention factor of 2–3; this assures the best resolution in the shortest time. In general, the selection of an isothermal column temperature depends on all of the other

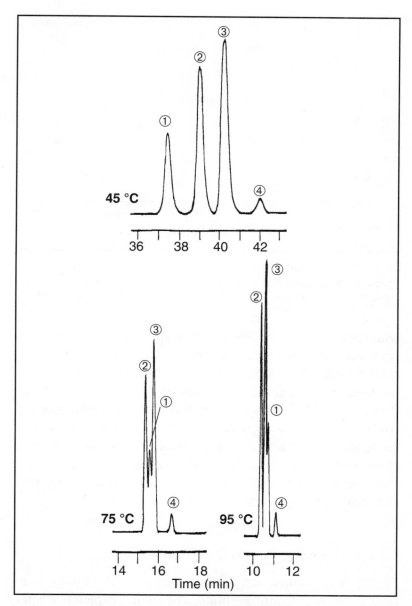

Figure 35. Separation of four hydrocarbons at different temperatures.
Column: *45-m x 0.50-mm i.d., support-coated open-tubular, coated with squalane stationary phase.* **Conditions:** *carrier gas, helium, 20 cm/s; oven, isothermal as indicated.* **Peak identification:** *(1) methyl-cyclohexane; (2) 2,5–dimethylhexane; (3) 2,4–dimethylhexane; (4) 2,2,3–trimethylpentane.*

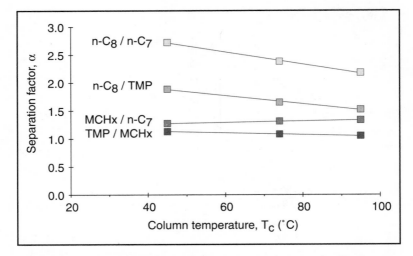

Figure 36. Relationship between the separation factor and column temperature for four peak pairs. Stationary phase: squalane. Sample: n–C_7 = n–heptane; n–C_8 = n–octane; MCHx = methylcyclohexane; TMP = 2,2,3–trimethylpentane.

conditions as well as on the sample itself. First, make a temperature-programmed run that covers the entire range of expected solutes. Then, if all solutes elute within a relatively narrow range of temperatures, choose a temperature close to the middle of that range and try isothermal elution. Some fine tuning of the isothermal temperature may be necessary if all solutes are not separated. If changing the isothermal temperature has little effect on the separation, then there is little difference in selectivity between the solutes. In that case is better to change to a different stationary phase than to hunt for the "right" temperature.

6.6.2 Programmed-Temperature Elution

In programmed-temperature operation, the column temperature increases during elution. The solutes experience what can be conceptualized as a series of short migration steps at incrementally increasing isothermal temperatures. It is as if the column were operated at one temperature for a short interval and then heated to the next for another interval, and then to the next, continuing until all peaks elute. The increasing temperatures cause the solutes' vapor pressures to increase. They spend less and less time in the stationary phase, and move down the column with increasing speeds, until they elute at their characteristic *retention temperature*, T_R (°C). The retention temperature depends on the rate of temperature increase as well as the other factors we have already discussed (stationary phase, column dimensions, and flow rate).

Programmed-temperature operation of a column can be described by the following general relationship:

$$T_t = T_0 + r_T (t - t_0) \qquad \text{eq.6.19}$$

where T_t is the column temperature at time t, T_0 is the initial column oven temperature that is held for the first t_0 time of the run, and r_T is the rate of temperature increase. Sometimes column temperature programming is interrupted by an intermediate isothermal period that is followed by additional temperature increase at a different rate. The program may be limited by an upper temperature (T_{max}) so that the last portion of the program is isothermal.

6.6.3 Temperature Program Optimization

In temperature program optimization, a number of variables have to be considered, and the significance of different peak pairs must be weighed since all peaks often cannot be completely separated. Harris and Habgood, in their fundamental study of programmed-temperature GC[81], pointed out that the most important parameter is the ratio r_T/F_{sp}, where r_T is the temperature program rate and F_{sp} is the *specific flow rate*. This is the average flow rate at column temperature ($\bar{F}$) per gram of stationary phase in the column:

$$F_{sp} = \frac{\bar{F}}{W_S} \qquad \text{eq.6.20}$$

where W_S is the amount of stationary phase. According to Harris and Habgood, the ratio:

$$\frac{r_T}{F_{sp}} = a \qquad \text{eq.6.21}$$

has an optimum which is independent of column type and other parameters. In other words, the ratio should be adjusted to keep a close to optimum. This rule permits us to investigate how the program rate should be selected for various columns. If we assume that the stationary-phase density is 1.0* and the film thickness is much less than the column i.d. ($d_f \ll d_c$), we can describe W_s by:

$$W_S = V_S = \frac{V_G}{\beta} \qquad \text{eq.6.22}$$

and

$$V_G = r_c^2 \pi L \qquad \text{eq.6.23}$$

where V_G and V_S are the volumes of the gas and stationary phases in the column. The average column flow rate ($\bar{F}$) is described in terms of column parameters by:

$$\bar{F} = \frac{\bar{u} \cdot d_c^2 \pi}{4} \qquad \text{eq.6.24}$$

and the phase ratio (β) by:

$$\beta = \frac{d_c}{4d_f} \qquad \text{eq.6.25}$$

Consolidation of these relationships yields the following expression for the specific flow rate:

$$F_{sp} = \frac{\bar{u} \cdot d_c}{4d_f} \qquad \text{eq.6.26}$$

---

* In practice, the stationary-phase density is about 0.8–0.9. Hence, this assumption results in a 10–20 % error in the calculations.

Substitution of this relationship into eq.6.21 leads to the following final relationship:

$$\frac{4r_T \cdot L \cdot d_f}{\bar{u} \cdot d_c} = a \qquad \text{eq.6.27}$$

This means that, in the case of a longer column, one should use a lower temperature program rate. The same rule holds for the stationary-phase film thickness; a lower program rate will be best in the case of thick stationary-phase films. For columns with a larger inner diameter or when operating at a higher average linear gas velocity, one would tend to choose a higher temperature program rate.

Proper selection of the temperature program also depends on the chemical and physical nature of the peak pairs which are the most difficult to separate. While the general guidelines given above for temperature program selection can suggest preferred relationships between the program rate and the column parameters, "fine tuning" of the conditions is best accomplished by trial and error. A good example of this type of optimization is found in the article by Johansen, Ettre, and Miller[82].

The elution process under temperature-programmed conditions may be thought of as occurring in a series of discrete column temperature increases. The relationship between temperature and the equilibrium coefficient given in eq.6.17 can be applied in order to calculate the distance each peak migrates through the column during the short time interval the column is at any one temperature. If the temperature intervals are made arbitrarily small, we can express the retention temperature in the following integral[81, 83]:

$$1 = \int_0^{t_R} \left[\frac{1}{t_M (k_t + 1)} \right] dt \qquad \text{eq.6.28}$$

where k_t is the retention factor as a function of the column temperature program. Since $K = \beta \cdot k$ (see eq.3.7 on page 31), we can write eq.6.17 in terms of the retention factor:

$$\log(\beta \cdot k_T) = \frac{A}{T_c} + B \qquad \text{eq.6.29}$$

Now, we can determine the constants A and B by measuring k for each solute at two isothermal temperatures. It is possible, then, to solve eq.6.28 for t_R. There are a number of commercially available computer programs that use these relationships to simulate the results of various temperature programs based on two or more runs under known conditions.

In addition to changing the separation factor and reducing analysis times, temperature programming has another important consequence: The carrier gas viscosity changes along with the column temperature. Figure 5 on page 25 plots carrier gas viscosity as a function of column temperature. Helium, for example, has a viscosity of 2.07×10^{-6} Pa·s at 50 °C and 2.86×10^{-6} Pa·s at 250 °C; a 38–% change. If the carrier gas inlet pressure is held constant during column temperature programming, the average linear gas velocity will decrease in proportion to the viscosity increase. If the linear velocity is close to optimum at the initial column temperature, then it could drop well below optimum at the final temperature. For this reason, the carrier gas velocity at the initial oven temperature is frequently set high enough above the optimum so that the final gas velocity will not fall below optimum. The effects of operation above the optimum gas velocity are minimized with helium or hydrogen carrier gas (see Section 6.4.2). Another possibility is programmed control of the inlet pressure so that the column flow rate (F_c) remains constant across the temperature program range[84, 85]. While this does not give a constant average linear gas velocity, it greatly reduces the changes in $\bar{u}$ with temperature.

❉ ❉ ❉

In this chapter, we have given a general overview of the interactions between some of the important variables in open-tubular GC. The references given here as well as the general books listed in Part IX give further detail.

Part VII

Inlets

Sample inlet systems form the first part of the chain of custody that passes the sample through a gas chromatograph. An inlet system conditions the sample so that utilization of the column's resolving power is maximized and peak area deviations are minimized. In this chapter, we explore sample introduction requirements for open-tubular columns and then examine the major sample introduction techniques that are used today. For more details on the various inlet systems, see refs. [86–88].

7.1 The Sample Introduction Process

Sample introduction adjusts a sample's physical state and amount from how it is presented for analysis to the state and amount required by the column for successful separation and elution. Sample passes from an external container, through the inlet system, and into the column. Successful sample introduction must meet certain requirements for the inlet, its operating mode, and the sample. The exact nature of the process depends upon the sample matrix and solute concentration. *Split* injection is usually required for concentrated samples in which solute concentrations exceed 0.1 %, since only a fraction of the sample vapor enters the column. *Splitless* injection is suitable for trace-level samples, since up to 98 % of the injected sample vapor passes into the column (the remainder is purged to the atmosphere). In *direct* injection, all of the sample vapor passes into the column after injection and evaporation in an intermediary inlet liner without a purging step; in *on-column* injection, liquid sample enters the column immediately from a syringe without exposure to other inlet components. These last three techniques all require some form of solvent management since large injected quantities of solvent interfere with the normal chromatographic process. Solutes may also have to be focused or trapped at the beginning of the column because the trace-level injection techniques may by themselves produce excessively large initial peak widths.

An inlet's performance is measured in terms of its ability to quantitatively and repeatably transfer all sample analytes into the column in equal proportions relative to the original sample. An inlet which does so is delivering linear performance. Inlet linearity means that the following criteria are fulfilled[89, 90]:

- ▼ The relative peak sizes in a sample mixture must be the same as either calculated values or the same as those found on the same detector without splitting.
- ▼ The peak areas must remain proportional to changing solute concentrations in the sample mixture.
- ▼ Relative peak sizes in a chromatogram should remain constant even when changing the analytical conditions such as temperature, split flow rate, column flow rate, and injected amount.

Naturally, any set of conditions must lie within reasonable operating limits. For example, one would not normally inject 100 µL of sample, or operate a vaporizing inlet system at temperatures far below the column temperature.

There are a number of ways that an inlet system deviates from linearity. In *mass discrimination,* analytes transfer into the column in relation to either their vapor pressures or molecular weights not just according to the sample composition. In *adsorption,* polar analytes are lost completely or partially on an inlet's inner surfaces. In *decomposition,* analyte thermal breakdown occurs in the heated inlet. All of these effects cause full or partial loss of some components, and may also add new peaks to a chromatogram from partial analyte decomposition. Matching sample requirements to inlet characteristics helps to control these effects.

7.1.1 Column Requirements

Open-tubular columns impose a set of sample requirements that are different than those of packed columns. Each column type has sample volume and amount limits, which if exceeded will degrade the subsequent chromatographic separation. Packed columns tolerate relatively large liquid sample volumes (1–5 µL) and solute amounts (up to 1 mg), which are on the same order as those conveniently injected from microliter syringes. Standard packed-column inlets are simple extensions of the column itself, with the same inner diameter and internal flow rate. The inlet is merely a sample vaporizing chamber that does not further condition the sample before it enters the column.

Open-tubular columns can accept only much smaller sample volumes and amounts. Smaller internal diameters (and therefore lower optimum flow rates) limit the overall sample vapor volume. Similarly, lower stationary-phase volumes limit the solute amounts that are tolerated before the chromatography seriously degrades. Excessive solute amounts distort peak shapes and shift retention times. Open-tubular columns tolerate about 0.1–3000 ng per component injected, depending on the column inner diameter and the stationary-phase film thickness. Table 16 on page 111 lists measured sample capacities for a number of columns. Sample capacities also depend on the solute and stationary-phase chemistry as well as on the column temperature and retention factor. The values listed in Table 16 represent practical working limits.

7.1.2 Extra-Column Contributions to Band Broadening

The total bandwidth of a peak (σ_{meas}^2) can be expressed in terms of contributions to the peak's measured *variance* from the column itself, plus contributions from sources outside the column. The column's contribution to the peak variance is equal to the square of the peak's standard deviation, σ_c^2 (see section 4.1.2 on page 43), a deviation that would be observed in the *absence* of any extra-column contributions. The *extra-column* contributions to band broadening, σ_e^2, contribute additively to the total measured peak variance:

$$\sigma_{meas}^2 = \sigma_c^2 + \sigma_e^2 \qquad \text{eq.7.1}$$

The *peak fidelity*, φ, expresses the relative contribution of extra-column variance to the peak's measured variance[91]:

$$\varphi = \sqrt{\sigma_c^2 / \sigma_{meas}^2} = \sqrt{\sigma_c^2 / (\sigma_c^2 + \sigma_e^2)} \qquad \text{eq.7.2}$$

Solving eq.7.2 for σ_e^2 with a 95-% peak fidelity, we obtain:

$$\sigma_e^2 \geq 0.108 \, \sigma_c^2 \qquad \text{eq.7.3}$$

The total amount of extra-column variance from the inlet, detector, connections, and so on, must be less than about 10 % of the variance from the column alone for a 95-% or better peak fidelity. Peaks will be significantly broadened by the extra-column contributions at lower peak fidelity levels, or the resolving power of the column will not be fully utilized.

Some injection techniques purposely deliver large sample bands into the column. In these cases, additional band sharpening techniques are needed to restore the required narrow sample bands (see the discussion on splitless injection in Section 7.3).

7.1.3 Inlet Requirements

In addition to meeting the column's requirements, the sample introduction process must also meet certain inlet system requirements. The sample amount is limited by the inlet system as well as by the column. Too much sample will "overload" an inlet, causing sample vapor to "flash" forward too far into the column and even back into the carrier gas supply pneumatics. Most inlet systems also have a minimum operating flow rate below which peaks become excessively broadened as well as a maximum flow above which retention time and area repeatability suffer. The inlet temperature is critical to obtaining the best performance; too cold a temperature may broaden peaks while one that is too high may cause solutes to break down. In addition to these considerations, some injection techniques have other specialized parameters that must be adjusted within certain limits.

7.2 Split Inlets

Split inlets are perhaps the most widely used injection systems for open-tubular columns. While a number of different designs have been described and implemented, virtually all split inlets in use today are similar to the concentric-tube type illustrated in Figure 37 on the next page. This design transfers analytes into the column accurately and reproducibly for a wide variety of sample types and concentrations. Split injection is not appropriate for trace-level samples since the analyte fraction entering the column may be below detection limits. In this case, a split inlet may be used for the related "splitless" injection technique, thereby transferring most of the sample into the column (see Section 7.3).

7.2.1 The Split Ratio

In conventional split injection, liquid sample is transferred from a microsyringe into a hot inlet liner where it

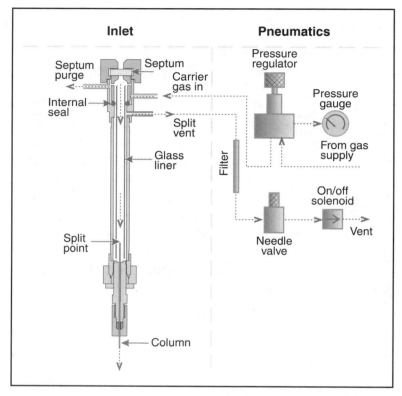

Figure 37. Typical split inlet system. Pressure-regulated pneumatics shown at right.

vaporizes and mixes with flowing carrier gas. A small fraction of the sample enters the column as the vapors pass by the column entrance (the *split point*), and the rest pass out of the inlet to vent. The relative sample amount entering the column is controlled by adjusting the vent and column flow rates. The ratio of the total carrier flow through the inlet to the column flow is called the *split ratio, s* (dimensionless):

$$s = \frac{F_v + F_a}{F_a} \qquad \text{eq.7.4}$$

where F_v and F_a are the vent and column flow rates respectively. These two flows must be expressed at the same temperature and pressure, usually room conditions. The column flow rate is controlled by the carrier gas pressure drop across the column. The split vent flow rate is controlled by a flow controller or needle valve placed in the inlet pneumatic lines.

Nominally, one percent of the sample enters the column when the split ratio is 100:1. The exact value of the split ratio may differ, however. Split injection is a dynamic process; sample vaporizes rapidly in a heated inlet, and the internal pressures and flows will change accordingly. The split ratio is measured while only pure carrier gas is passing through the inlet. Sample vapors have a significantly different viscosity than pure carrier gas, so that the column and inlet flow rates at the split point may change during injection. Furthermore, sample vapors may pass out of the split vent and through the split flow controlling device before the injection is complete. The split vent flow may then change during injection, affecting the split ratio. Most split inlet systems employ a *buffer volume*[92, 93] and/or an adsorbent trap[94] between the split vent and the flow-controlling device so that sample vapors do not affect the split vent flow until after sample transfer into the column is complete.

Split inlet systems should be recalibrated for quantitative analysis whenever the split vent flow, column flow, or inlet temperature changes. This includes changing the column oven temperature at injection since the column flow is affected. The split ratio should never be entered into a data handling system for inclusion in quantitative calculations but should always be recorded along with other instrument conditions. The split ratio is only an approximation of the actual split; its value is usually rounded after calculation.

Example: Calculating the split ratio. A 25-m x 0.25-mm i.d. column is operated with 12.7 psig helium carrier gas in a GC oven at 100 °C. The column outlet flow corrected to room conditions is 1.19 mL/min (see example on page 24). The split vent flow rate is measured at 137 mL/min. The split ratio is calculated from the flow rates:

$$s = \frac{137 + 1.19}{1.19} = 116:1$$

The value of 116:1 normally would be recorded as 120:1.

The split vent flow must be greater than about 20 mL/min: Sample transfer into the column is retarded at lower vent flows, which may cause extra band broadening and reduced utilization of column resolving power. The vent flow should be limited to about 350 mL/min. Pressure drops through the pneumatic supply tubing at very high vent flows may cause the measured split ratio to differ from the actual ratio, and the pressure at the column inlet may be different than indicated on the instrument head pressure gauge or readout.

7.2.2 Split Injection Linearity

Vaporization occurs rapidly in conventional hot split injection, forming a relatively large amount of gas that expands to fill the available space. One microliter of solvent will generate between 100 and 1000 μL of vapor, depending on the inlet temperature and pressure as well as the solvent's molecular weight and density. A large internal diameter inlet liner ($\approx$ 4–mm i.d.) provides about 1000 μL for sample expansion. A smaller liner ($\approx$ 2–mm i.d.) has only about a 250–μL volume and can tolerate only very small injected volumes. The larger liner is preferred in most cases, although the smaller liner will give sharper peaks at lower split vent flows. Pressure in the inlet liner can momentarily increase 2–5 times the set pressure during injection, disrupting both the column and vent flow rates. Sample and solvent vapors can be driven out of the liner and even back into the carrier gas supply lines. They may diffuse slowly back into the inlet system and reappear during subsequent injections. A *septum purge* vent[95] (see Figure 37 on page 132) removes sample vapor from the area above the inlet liner, including any volatile material outgassing from the septum. The septum purge helps to maintain a sharp solvent peak profile at higher injected amounts, although it cannot alleviate some of the other side effects.

Sample that is lost out of the liner due to this vapor overflow effect is not available to the column, however. This loss may affect the linearity of the transfer process from inlet to column, so that eluted peak areas are not representative of the original sample (mass discrimination). The more-volatile sample components tend to be lost preferentially out of the liner so that the portion of sample actually entering the column may be enriched in the less-volatile sample components.

Disruption of flow in the liner from rapid vaporization and subsequent momentary high internal pressures may result in further nonrepresentative sampling into the column as well as poor repeatability. For these reasons, injected sample amounts in conventional split injection should be limited to less than a microliter. If peak areas are small, the split vent flow should be decreased (causing more sample to enter the column), instead of increasing the injected volume. Splitless injection (Section 7.3) or another of the trace-level techniques should be used instead if the split vent flow is already under about 20 mL/min.

While vapor overflow from the inlet liner may cause a loss of the more volatile components, syringe handling techniques may produce the opposite effect: an enrichment of the volatiles. Manual syringe injection can take from 1 to 5 seconds, depending on the individual. The syringe needle is heated by the inlet during its residence there. It takes about 1 second to bring the needle up to a nominal inlet temperature of 250 °C. Sample present in the syringe needle before or after the plunger is depressed will begin to vaporize, or *fractionate* into the inlet; the more-volatile components will evaporate first, while the less-volatile ones will take longer to evaporate. The volatile components will be enriched inside the inlet if the plunger is depressed or the syringe needle is withdrawn from the inlet before all of the sample in the needle has evaporated. Therefore, the portion of sample entering the column will also have an excess of volatiles relative to the original sample composition.

This syringe needle discrimination effect has led to the development of a number of specialized syringe handling techniques including *hot needle* injection, *sandwich* injection, and *high-speed* injection[87]. In hot needle injection, sample is withdrawn completely into the syringe barrel before injection. The operator waits a few seconds after placing the needle in the inlet before depressing the plunger, and then pauses after injection to allow the needle to heat up again and the remaining sample to evaporate from the needle. In sandwich injection, the syringe is loaded first with about 1 µL of pure solvent, then with an air bubble, and finally with the sample. The additional solvent plug displaces residual sample from the needle into the inlet, leaving only the pure solvent behind in the needle. Any additional evaporation from the needle is with solvent only. High-speed injection uses a mechanical autosampler that injects sample in under 0.5 s. The entire injection cycle takes place so rapidly that the syringe needle never heats significantly, and sample fractionation from the needle is effectively suppressed. Perhaps the best solution to this problem, however, is to switch to a cold or temperature-programmed injection technique (see Sections 7.5 and 7.6).

7.2.3 Liner Packing

The split liner is usually packed with deactivated (silanized) glass wool or glass beads to facilitate sample vaporization and mixing with carrier gas[96]. Liquid sample often does not evaporate immediately upon ejection from the syringe. It may form an aerosol cloud or larger droplets that can move past the column tip and out towards the split vent. The droplets will partially evaporate as they move through the inlet, concentrating the less volatile components which are no longer available to the column as they are swept away from the column entrance. Adding packing to the liner helps to better evaporate the sample before it reaches the column. Liner packing also wipes off any droplets clinging to the syringe needle, which otherwise would be pulled up and left on the inner septum surface as the needle is withdrawn. Both of these effects can lead to nonlinear splitting in the absence of liner packing.

7.2.4 Adsorption and Decomposition

Exposure to heated inlet surfaces can cause adsorption and thermal or catalytic decomposition of sensitive samples. These problems may be enhanced by the presence of high surface area liner packings. Careful liner and packing deactivation by silanization reduces the available active silanol sites on their surfaces. Replacing borosilicate glass liners and glass wool packings with quartz linersand quartz wool eliminates catalytically active metal sites present in conventional glasses.

7.3 Splitless Injection

Split injection is inappropriate for trace-level samples that do not give sufficient area counts at low split ratios. Splitless injection offers an alternative that transfers nearly all of the injected sample into the column[87]. The choice depends not only on the sample, but also on the detector. High-sensitivity devices such as an electron-capture detector (ECD) can deliver acceptable results with more dilute samples than a less sensitive flame-ionization detector (FID).

Transferring large sample amounts into the column produces severe overloading and distortion of the major components. This is of little concern since the solvent generally is the only major component in a trace-level sample. However, the presence of a large solvent peak may create additional problems for the separation of the peaks of interest. Early peak retention times may shift, and all peaks may be distorted and broadened. Controlling these effects can be a major challenge with splitless injection as well as with the other trace-level sampling techniques.

Splitless injection is performed in a split inlet equipped with a valve that turns on and off the split vent flow passing through the inlet liner. This is usually an automatic electric solenoid activated by the GC's software at programmable times after injection. When the vent flow is interrupted, the only flow through the inlet liner is into the column itself. During the off period sample flows only into the column; this time period is called the *splitless sampling time*.

7.3.1 Recovering Peak Shape

When sample is injected, its vapors begin to pass into the column. Splitless sample transfer into the column takes considerably longer than in split injection, because the liner flow rate is reduced to the column flow rate. It will take about 15 seconds to transfer over 90 % of the sample vapor from a 2–mm i.d. liner at a column inlet flow of 2 mL/min (at the inlet pressure and temperature). This time increases to over one minute with a 4–mm i.d. liner; it is important not to turn on the split vent flow too soon after injection. The appropriate splitless sampling time should be determined experimentally for specific samples and chromatographic conditions by measuring area counts with increasing sampling times until no further area increases are observed.

Splitless sample transfer times are far too long for acceptable peak shapes, especially early in the chromatogram. Peaks would be severely broadened without additional band sharpening after injection. *Cold trapping* on the column is one peak shape recovery technique used with splitless injection. The column oven is kept at a reduced temperature, at least 30 °C below the elution temperature of the first peak of interest. Analyte vapor encounters the cool column and experiences a high retention factor: It is trapped in the stationary phase and does not migrate down the column. Peaks condense into a very narrow band at the juncture between the column and the inlet. After injection is complete, the oven is temperature-programmed, and the solutes elute with normal peak shapes.

7.3.2 The Solvent Effect

Solutes which are close to the solvent peak may experience an additional band-sharpening process. The solvent presents a severely overloaded peak shape which increases gradually from beginning to end and then drops off very quickly. The rapid decrease is due both to the overloaded peak shape as well as the quick inlet purging that occurs upon opening the split vent after sample transfer into the column. The solvent acts as a very thick, temporary stationary-phase coating and imparts a very high effective retention factor on peaks trapped therein. Volatile peaks that are not cold-trapped will tend to concentrate at the back of the solvent peak (the thickest section) and take on a narrow profile. It is possible, for example, to separate traces of isooctane in normal octane using this *solvent effect*[97, 98]. The initial column oven temperature should be close to, but lower than, the solvent boiling point to ensure a strong solvent effect. The retention of solvent-affected peaks tends to shift towards later times, depending on the solvent amount and the temperature, so that their retention times may depend on the injected volume.

7.3.3 Use of a Retention Gap

In some instances, the solvent vapors may condense inside the column. For larger sample amounts there may be enough solvent present to form droplets in the column. Solvent droplets tend to coalesce into blockages, which are propelled rapidly by carrier gas flow down the column until they break up. The droplets may coalesce again, moving down the column in steps until one or more meters are coated with liquid solvent. This situation is called *solvent flooding,* and it can be especially detrimental to middle- and late-eluting peak shapes. It can also disrupt the stationary-phase coating and shorten column lifetime. Solutes dissolved in the solvent are swept along the column as well, breaking up into a series of ragged bands. Unless something is done to restore their peak shapes, they will elute from the column in this condition. One possibility is to reduce the amount of solvent by concentrating the sample before injection. Alternatively, a higher initial column temperature will suppress solvent condensation during sampling.

Another technique employs a 1–3 m piece of uncoated column tubing installed between the inlet and the column. Solvent flooding is allowed to progress inside this *retention gap*[99, 100], but is contained therein. Once spread out inside the retention gap, solutes will quickly migrate towards the column as the solvent evaporates under the influence of increasing oven temperatures. Solutes will be trapped when they encounter the stationary phase at the beginning of the analytical column itself, whereas solvent will pass through the analytical column in the vapor phase. The name *retention gap* refers to the difference in retention power between the uncoated precolumn and the analytical column. While they can move through the precolumn at relatively low temperatures, solutes cannot overcome the increased retention of the analytical column, and so they are trapped there until the oven temperature rises enough to permit them to migrate farther down the column.

Splitless injection can accentuate problems with sample adsorption and decomposition. Solutes are exposed to the heated inner inlet surfaces for an extended time (compared to split injection), and there is more opportunity for thermal and catalytic decomposition. Careful inlet liner deactivation combined with judicious selection of inlet temperatures, splitless sampling times, and inlet flow rates will help minimize these effects. Many sensitive samples will require cold on-column or programmed-temperature injection for the best results (see Section 7.5 or Section 7.6, respectively).

7.4 Direct Inlets

A direct inlet for open-tubular columns (see Figure 38) most closely resembles a conventional packed-column inlet. All of the carrier gas flow goes either to the column or to a septum purge: There is no split vent. Sample is deposited in an inlet liner where it vaporizes, expands, and passes directly into the column, making direct inlets suitable for trace analysis. The inlet operates at the column flow rate; there is no split vent flow to clear out residual sample vapors after injection. For this reason, direct inlets generally are used with columns that operate efficiently at flows of 5 mL/min or higher such as wide-bore columns having $d_c \geq 0.53$ mm. The lower flow rates of narrower i.d. columns would not sweep out the inlet rapidly enough, and would lead to broadened or tailing peaks.

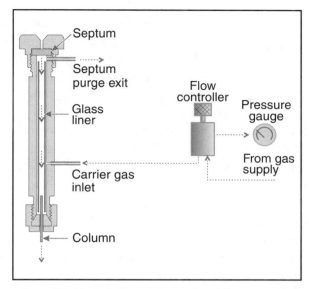

Figure 38. Typical direct inlet system. Flow-controlled pneumatics shown at right.

In contrast to splitless injection on narrow-bore columns, direct injection on some thick-film wide-bore columns is suitable for both percent-level and trace-level samples. The higher sample capacity of wide-bore columns (1–3 µg, see Table 16 on page 111) allows successful separation of higher concentration samples that would not be tolerated by narrow-bore columns. This compatibility permits the substitution of wide-bore columns for packed columns in many instances. Packed-column inlets are routinely adapted to direct injection onto wide-bore columns. The conversion enables similar resolution in less time with wide-bore columns operating at higher flow rates (over 10 mL/min) or better resolution in about the same analysis time when operating wide-bore columns closer to optimum flow rates (under 10 mL/min). See Section 6.1 on page 92 for a comparison of packed and wide-bore columns.

Sample introduction techniques using direct inlets are similar to those for splitless injection. Injection volumes must be limited so that sample vapor does not overfill the inlet liner and escape into the septum purge or the pneumatic supply lines. The effects of a too large injection on peak shapes are more evident since there is no post injection purging step as found in splitless injection. Solvent peak shapes can be managed with the solvent effect, although it may be more difficult to obtain a strong solvent effect on a high capacity thick-film column. A retention gap also helps control the side effects of solvent flooding.

Solute peaks can be narrowed with the solvent effect as well as with cold trapping, although these effects are not as critical since wide-bore columns generate peaks that are wider than those from narrow-bore columns. Direct inlets should be packed with glass wool, not for sample dispersion in the carrier gas but for more reproducible injection volumes; the glass wool will remove liquid droplets clinging to the syringe needle.

Direct inlets are subject to the same difficulties with sensitive peak adsorption and decomposition as split inlets. Sample residence times in direct inlets are longer than those found in split injection, but generally shorter than in splitless injection, and the effects of sample–liner interaction are intermediate between split and splitless injection. Obtaining the best results requires the same careful liner and packing deactivation as for splitless injection.

7.5 On-Column Inlets

In principle, on-column inlets are the simplest type of open-tubular column inlet. Sample is deposited directly into the column or its extension (both of which are held at a low injection temperature), usually the same as the initial oven program temperature. Sample does not contact the inner inlet surfaces in cold on-column injection and is not subject to the same deleterious effects as when deposited into a hot inlet liner. Solute adsorption and decomposition are greatly reduced. Furthermore, the problems stemming from rapid solvent expansion and syringe needle heating are eliminated if the injection temperature is close to or less than the solvent's boiling point. Sample is essentially pipetted into the column from the syringe.

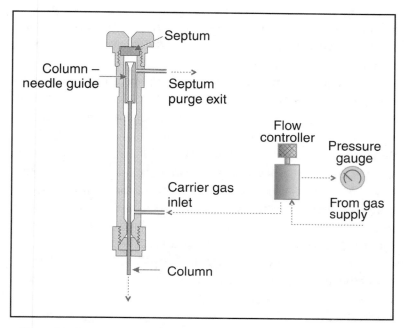

Figure 39. Typical on-column inlet system. Flow-controlled pneumatics shown at right.

The inlet and column are not heated to begin the elution process until the sample has been transferred into the column. While it is possible to perform on-column injection with a hot inlet, sample sizes are very limited due to the small expansion volume (the column itself). "Hot" on-column injection is not in widespread use.

The first cold on-column inlets were arranged so that the syringe tip penetrated the column to a point inside the column oven, thus ensuring injection at the oven temperature[101]. An air cooling stream was directed across the inlet, with additional air across the first few centimeters of column in the oven to help dissipate residual heat from the previous injection. This "secondary" cooling system improved analytical accuracy and repeatability and reduced the cooldown time between runs. More recently, on-column inlets have incorporated independent inlet temperature control. The inlet temperature is controlled at or slightly above the oven temperature by the GC microprocessor, ensuring that the entire column is heated uniformly. Some on-column inlets use cryogenic liquids for more rapid cool-down times between runs.

On-column injection has been applied to columns with inner diameters of 0.18 mm and larger. Column diameters under 0.53 mm require special syringe needles made of fused-silica or narrow steel tubing which are too thin to be able to penetrate a conventional solid septum. In this case, the inlet uses a valve or special septum instead, hence the term "septum-less" on-column injection. On-column injection onto narrow-bore columns also is made difficult by sample particulates and nonvolatile residues normally found in "real world" samples. Careful sample cleanup is often required.

On-column injection into 0.53-mm and larger columns is possible with standard syringes that use 0.47-mm o.d. needles. These needles are robust enough to penetrate standard septa without damage, and they tolerate the added stress of autosampler injection quite well. For narrower bore columns, a short piece of 0.53-mm fused-silica column can connect the inlet to the column with a zero-dead-volume union in the column oven. The 0.53-mm i.d. tubing acts as a precolumn that traps sample residue, and it easily can be replaced without compromising the analytical (narrow-bore) column itself

A longer precolumn is installed when a retention gap is required; cold on-column injection is more prone to solvent flooding effects than hot injection techniques because the liquid solvent enters the column directly. In the other techniques the solvent enters the column as a vapor that must recondense before flooding can occur.

7.6 Programmed-Temperature Injection

The concept of independent inlet temperature control has been applied to split and splitless injection as well. In a *programmed-temperature* inlet, the injector temperature also is controlled by a separate heating and cooling arrangement. A programmed-temperature inlet is also called a *programmed-temperature vaporizer* (PTV)[102–104]. A glass or quartz liner receives sample from the syringe at a reduced initial temperature equal to or less than the column oven initial temperature. Sample transfer from the syringe into the inlet is free of hot injection effects, although some liner packing is required to remove any sample droplets that might cling to the syringe needle upon withdrawal. Then, after syringe needle withdrawal the inlet temperature is rapidly programmed up to the final oven program temperature or slightly higher at rates as high as 1200 °C/min. The high inlet temperature program rate ensures sample vaporization rapid enough to generate narrow initial peak profiles but not so rapid as to form excessive sample vapor or too high an inlet pressure. As a result, programmed-temperature injection does not suffer from the side effects of explosive sample vaporization, in contrast to hot split or splitless injection, and it is relatively free of mass discrimination and other nonlinear splitting effects.

The controlled heating of a programmed-temperature inlet significantly reduces sample adsorption and decomposition because not all solutes are exposed to excessively high temperatures. In a conventional hot split inlet, the temperature must be hot enough to vaporize all sample components rapidly, including those that do not require such high temperatures. These more volatile constituents are exposed to temperatures much higher than necessary. In programmed-temperature split injection, the components vaporize into the carrier gas stream and enter the column as the inlet is heating. They leave the inlet and are not exposed to higher temperatures than required for their vaporization. Programmed-temperature split injection has been applied to many different sensitive samples including drugs and herbicides. However, programmed-temperature splitless injection holds sample in the inlet during the heating period for a longer time and does not benefit as much as programmed-temperature split injection from lower solute vaporization temperatures.

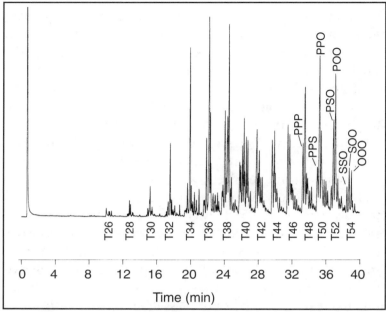

Figure 40. Separation of triglycerides in butter fat. **Column:** *25-m, 0.25-mm i.d., 0.12-μm film 65%-phenyl 35%-methylsilicone high-temperature phase on fused-silica.* **Conditions:***H_2 carrier gas at 60 cm/s; oven, 200 °C initial, 4 °C/min to 380 °C; inlet, programmed-temperature split, 50 °C initial, ballistic heating to 400 °C in 60 s.* **Sample:** *1 μL of 1% butter solution in octane. Fatty acid side chains: P = palmitin (C16:0), S = stearin (C18:0), O = olein (C18:1); T-numbers indicate total number of carbons in the side-chains. The baseline of a blank run (no injection) was subtracted from the chromatogram prior to plotting[105].*

Programmed-temperature split injection is often applied to high molecular weight samples, such as the high-boiling wax sample shown in Figure 26 on page 85. Another PTV application is the analysis of triglycerides. Figure 40 shows the separation of triglycerides present in butter fat. Triglycerides are subject to both mass discrimination effects due to their high molecular weights and to thermal decomposition at elevated inlet temperatures due primarily to exposure to the heated steel syringe needle. Both effects are largely absent with programmed-temperature injection because the syringe needle is not heated.

High molecular weight residues left behind in the inlet can produce additional mass discrimination effects in splitless programmed-temperature injection[104]. High molecular weight sample components may take a considerable time to evaporate inside the inlet in the presence of such residues[106]. Splitless injection may discriminate against such components if the vent flow is resumed before all of the sample has had a chance to evaporate and enter the column. On-column injection can produce similar effects[88].

7.7 Selecting an Inlet

Selection of an inlet for a specific sampling task usually is dictated first by the sample itself and second by available injection hardware. Table 17 on the next page presents some general guidelines. Many samples can be handled successfully by more than one inlet and column. If there are only a few components then almost any inlet combined with the right column can accomplish the separation. In this case, a wide-bore column will function with split, splitless, direct, or on-column injection*. Column selection is the more critical choice since one must rely on stationary-phase selectivity and not so much on brute-force resolution (see Section 5.6 on page 88). Wide-bore column solute capacity may have to be balanced against maximum or minimum oven temperatures. A thick-film high-capacity column may require too-high elution temperatures, or conversely a thin-film low retention column may not have sufficient sample capacity for splitless, direct, or on-column sampling. More concentrated samples may require dilution if an inlet splitter is not available.

Narrow-bore columns are needed for more complex samples. The right stationary-phase selectivity must be combined with sufficient resolving power. As the column i.d. is reduced below 0.53 mm, direct injection (using an adapted packed-col-

---

* A wide-bore column (0.53-mm i.d. or higher) must operate at a sufficiently high inlet pressure to ensure normal operation in split or splitless injection. Pressures above 5 psig are usually high enough.

Injection Technique	Column	Sample
Split	All	> 10 ppm (FID) > 100 ppb (ECD)
Splitless	Thick-film ($d_f > 1$ μm)	< 1000 ppm
	Thin-film ($d_f < 1$ μm)	< 100 ppm
Direct	$d_c \geq 0.53$ mm	Same as splitless.
On-column	$d_c \geq 0.25$ mm	Compounds sensitive to decomposition / adsorption. Same concentrations as splitless.
Programmed-temperature split or splitless	All	Compounds sensitive to decomposition / adsorption. Same concentrations as split or splitless samples.

Table 17. *General guidelines for matching injection techniques, samples, and columns.*

umn inlet) is no longer an acceptable choice. Now one must employ an inlet designed for open-tubular columns. An on-column inlet is the simplest choice, but its applications are limited with higher solute concentrations or with columns under 0.20-mm i.d. A split inlet with splitless capability becomes the most universally applicable choice for the narrower column inner diameters.

Special sample considerations are also important. Samples that are prone to inlet-related side effects such as mass discrimination, adsorption, or decomposition may require the special handling of a cold on-column or temperature-programmed inlet. The added complexity and particular operating technique of the latter make it a last resort when all else fails.

Part VIII

Detectors

The last piece of hardware that the sample encounters on its path through the GC is the detector. Simply put, the detector generates a signal in response to the presence of one or more sample components at the column outlet. The signal is then conditioned as required for transmittal to a data handling device, such as a computer-based data handling system or perhaps a simple chart recorder. The data handling device converts the incoming signal into a readable and archivable form such as a chromatogram plot, a printed report of peak areas, or representative calculated amounts. In this chapter, we will discuss the fundamentals of GC detectors for open-tubular columns.*

---

* For more information on the detectors used in gas chromatography, including the pertinent references, see the books specially devoted to this subject [107,108].

8.1 Detector Types

GC detectors fall into two broad groups: those that measure changes in the bulk properties of the carrier stream, such as thermal conductivity or density, and those that chemically respond to the presence of solute molecules, for example, by ionization or light emission. *General-purpose* detectors respond to nearly all solutes without regard to chemical functionality. *Selective* detectors respond only to specific chemical functionalities in solutes, such as halogenated or aromatic groups. Within these broad groups, *mass-flow dependent* detectors respond according to the mass flow or flux of solute passing through, while *concentration-dependent* detectors give a signal according to the solute concentration in the detector's active cell area.

Detector response to solutes is characterized by a number of quantities. *Sensitivity* expresses the magnitude of the basic detector response to an amount of incoming solute. *Detectivity* (for mass-flow dependent detectors) and *minimum detectable quantity* (MDQ, for concentration-dependent detectors) measure the smallest amount of solute that produces a measurable detector response. The response is defined as an output signal that is two times the output noise level. *Selectivity* is the ratio of response for one type of chemical functionality versus another, for example, for phosphorus to carbon in a nitrogen-phosphorus detector (NPD). *Dynamic range* is determined by the range of amounts or concentrations across which the detector sensitivity remains linear within a nominal tolerance such as ±5 %.

8.2 Detector Requirements for Open-Tubular Columns

All detectors impose a finite volume and response time onto the chromatogram emerging from the column. Detector fidelity is affected if either of these are too large, thereby distorting peak shapes, reducing the observed resolution, and potentially affecting response. Peak dilution inside the detector lowers concentration dependent detector sensitivity when the internal volume is large, relative to the eluted-peak volume. Slow response times also cause peak tailing and may decrease peak resolution in extreme cases.

8.2.1 Volume and Time Constant

Open-tubular columns produce chromatograms with faster peaks that occupy less volume than typical packed-column peaks. Detectors designed for packed-column use may not be suitable for open-tubular columns without modification. *Makeup gas* is frequently added to such detectors along with the column effluent. Makeup gas flow brings the internal detector flow rate up to packed-column levels and sweeps detector internal volumes more efficiently than low open-tubular column carrier gas flow rates. Of course, makeup gas reduces concentration-dependent detector sensitivity in direct proportion to the flow increase. Many modern detectors are designed for open-tubular column use and require no makeup gas.

Detector response times are not generally a problem as long as eluted-peak widths-at-base remain greater than about 0.5 s. Some detectors have selectable *time constants*, or τ_d (s), ranging from about 50 ms up to 1 s. For general use, $\tau_d \approx 200$–300 ms is suitable. Longer time constants may reduce baseline noise, giving a slight improvement in detectivity or MDQ. This effect is not useful for most open-tubular columns, however, because the time constant increase may cause unacceptable peak tailing. A too-fast time constant will increase the detector noise level without significantly improving the peak signal level or peak fidelity. Very narrow bore columns ($d_f \leq 0.100$ μm) require time constants of 50 ms or less because of the extremely narrow peaks they can generate.

8.3 Flame-Ionization Detector

The *flame-ionization detector* (FID) is a mass-flow sensitive ionization detector that responds to organic compounds. The FID is considered a general-purpose detector because most of the compounds analyzed by GC are organics, but this is not quite true: It does not respond significantly to water or atmospheric inorganic gases. Its good detectivity ($\approx 2 \times 10^{-12}$ g-carbon/s) make it the detector of choice in many situations. It is particularly well suited for open-tubular columns because it has an inherently fast speed of response (under 50 ms) and a low effective internal volume.

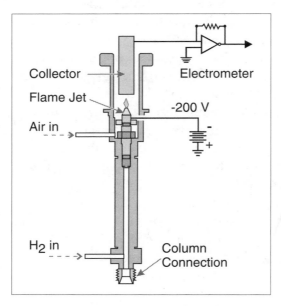

Figure 41. Typical flame-ionization detector.

A FID is illustrated in Figure 41. Column effluent enters the detector and mixes with hydrogen combustion gas plus makeup gas (if required) in the lower area. The gas mixture passes through a flame jet and combusts in an excess of air. Organics are decomposed and ionized as they enter the flame.

An electrical field repels electrons into a collector electrode, and the resulting current is amplified and converted to a voltage by an electrometer.

An FID may not be suitable in all situations where a general purpose detector is required because the hydrogen combustion gas and flame may be inappropriate in hazardous areas. In such cases, another general purpose detector such as the TCD may be a better choice.

8.4 Thermal-Conductivity Detector

The *thermal-conductivity detector* (TCD) is a concentration-dependent bulk-property detector that responds to all compounds with thermal conductivities different than the carrier gas. The TCD is most frequently used for gas analysis with adsorption or molecular sieve columns, but it also finds many applications where the extra air and hydrogen supplies of an FID are not available: TCDs require only the carrier gas for operation. They have an MDQ of around 10 ppm of C_9 in the TCD cell, which is significantly less sensitive than an FID. They are not appropriate for trace-level analyses.

Thermal-conductivity detectors contain one or two heated filaments that are suspended inside the carrier gas stream eluting from the column (the "sample" side) and one or two more filaments suspended inside a pure carrier gas stream (the "reference" side) flowing at the same rate. Many different geometries and flow paths are possible, but it is important to limit the TCD cell volume for open-tubular column use. Figure 42 on the next page illustrates a flow-through arrangement that has four filaments, two each for the sample and the reference sides. The filaments are suspended across the cell so that the volume occupied by electrical connections is minimized. The filaments are heated by an electrical current and reach a constant temperature when the chromatogram is at baseline. Heat is lost from the filaments into the carrier gas by conduction and lost onto the cell walls by radiation. Practically all solutes have

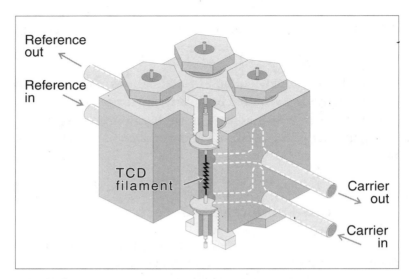

Figure 42. Typical thermal-conductivity detector.

thermal conductivities less than pure carrier gas, so filament temperatures increase as peaks pass over them. The change in filament temperature produces a change in filament resistance. The TCD electronics maintain either a *constant current* through the filaments, allowing the temperature to rise, or act to keep the filaments at a *constant temperature*, reducing the current as required. In either case, an output voltage is produced in proportion to the change in the detector electronics' operating conditions. TCD filaments can burn out if operated too hot or if exposed to excessive oxygen from air leaks. Most TCD electronics can detect dangerously high filament temperatures, such as those that occur during very large peaks or when there is a leak, and will shut down TCD operation to protect them.

Peaks are exposed to a total cell volume of 100–150 µL in a typical TCD for open-tubular columns. This volume is sufficiently low to avoid significant extra band broadening on 0.53-mm i.d. columns at or above optimum flow rates (3–5 mL/min). Narrower i.d. columns operating at lower flow rates require makeup gas, which will dilute the column eluent and reduce detector response.

TCD response depends on the difference between the carrier gas and solute thermal conductivities as well as on the temperature differential between the filaments and the TCD cell housing. Helium carrier gas is the best choice for most open-tubular columns; it has the highest thermal conductivity of the common carrier gases (except for hydrogen), and it gives better column performance overall. Nitrogen also may be used with a TCD, but its lower thermal conductivity limits filament currents and gives a lower response to eluting peaks.

8.5 Electron-Capture Detector

Both the FID and TCD are general-purpose detectors that respond to a wide range of substances. Many target analytes contain specific heteroatoms or chemical functionalities to which selective detectors respond strongly while giving little or no response to analytes without the requisite properties. The *electron-capture detector* (ECD) is a selective concentration-dependent detector that responds strongly to halogenated molecules. The ECD is used primarily for trace-level analysis of halocarbons, arochlors, and pesticides because it has a very low MDQ (50×10^{-15} g for lindane) yet is highly selective and does not respond significantly to hydrocarbons at analytical levels (< 1 µg).

The ECD contains an active β-particle source, usually ^{63}Ni, which ionizes nitrogen carrier gas or a 5-% methane-in-argon carrier gas mixture, forming electrons (see Figure 43 on the next page). In a *constant-current pulsed-source* ECD, the electrons are gated onto a collector electrode by pulsing the electrode to a positive potential at a *base frequency* of around 250–500 Hz, creating a background current. Analytes enter the ECD cell and capture some of the electrons, reducing the background current. The ECD electronics increase the pulse frequency to maintain a constant background current, converting the required pulse frequency into the output voltage. Makeup gas is usually required with an ECD because the internal volumes tend to be

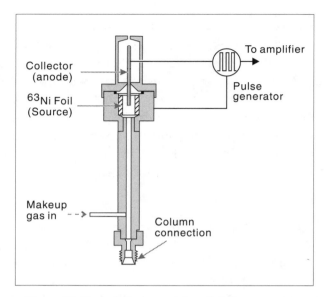

Figure 43. Typical electron-capture detector.

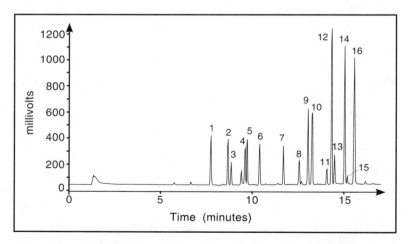

Figure 44. Analysis of pesticides with an electron-capture detector. Column: 30-m x 0.53-mm i.d. x 0.5-μm film dimethyl-35%-diphenyl silicone on fused silica. Conditions: oven, 150 °C, 8 °C/min to 270 °C; helium carrier, 8 mL/min; direct injection, 300 °C; 1 μL of 100–600 pg each component in methanol. Peak identification: (1) α-BHC; (2) γ-BHC; (3) β-BHC; (4) heptachlor; (5) δ-BHC; (6) aldrin; (7) heptachlor epoxide; (8) endosulfan-I; (9) p,p'-DDE; (10) endrin; (11) p,p'-DDD; (13) endosulfan-II; (14) p,p'-DDT; (15) endrin aldehyde; (16) endosulfan sulfate.

large for open-tubular column requirements. When helium is used as the carrier gas, nitrogen or 5-% methane-in-argon is added as the makeup gas, also providing the "working" ionizable gas inside the detector cell.

Electron-capture detectors are susceptible to signal suppression, or *quenching*, from large levels of hydrocarbons or water in the detector cell. Oxygen or halogenated solvent contamination in the carrier gas causes increased background signal and noise levels, resulting in reduced detector sensitivity. For these reasons, stringent sample and carrier gas cleanup are required for successful ECD operation.

8.6 Other Selective Detectors

A number of other selective devices are used for detection of specific analytes. Among them are the *mass-spectrometric detector* (MSD), the *nitrogen-phosphorous detector* (NPD or TSD), the *photoionization detector* (PID), the *electrolytic-conductivity detector* (ElCD), and the *flame-photometric detector* (FPD). Each of these are summarized below.

8.6.1 Mass-Spectrometric Detector

The MSD consists of a quadrupole or ion trap (ITD) mass spectrometer with limited-range unit mass resolution ($m/e < 650$ amu) connected to the column exit. Most MSDs fit on the benchtop and are not fully-fledged high-resolution mass spectrometers. MSDs produce *mass chromatograms* for single or multiple ions, as well as a *total ion chromatogram*. A MSD can be tuned to specific molecular ions or to fragments characteristic of various functional groups. MSD-generated spectra can be searched in a number of spectral libraries for positive peak identification.

The column may interface to a MSD in a number of ways. A *direct interface* simply places the column end inside the ionization source. Lower inlet pressures are required for normal carrier gas linear velocities since the column end is at vacuum.

The column flow rate is limited by the vacuum system, since carrier gas must be pumped away to maintain the required low ionization source pressure. Most MSD systems have difficulty pumping more than about 2 mL/min, so the column internal diameter is usually 0.32 mm or less. An *open split* interface keeps the column end near room pressure; column effluent is mixed with makeup gas and conducted to the MSD source through a fixed restrictor tube. The restrictor inner diameter and length are selected to maintain a constant flow close to 1 mL/min. Any flow in excess of the restrictor flow is vented to the atmosphere. Normal column pressure drops are employed, but some of the column effluent is lost, giving somewhat reduced sensitivity.

8.6.2 Nitrogen-Phophorus Detector

The NPD operates by catalytic thermionic reaction of nitrogen and phosphorous containing organic solutes on or near the surface of a hot rubidium- or cesium-containing ceramic bead. The NPD is sometimes called the *thermionic-specific detector* (TSD). The catalytic reaction produces electrons which are collected and amplified in much the same manner as an FID. A typical NPD has a detectivity of 5×10^{-13} g/s for nitrogen. The NPD finds application in pesticide and herbicide analysis as well as for drug screening. The separation of a drug mixture with nitrogen-phosphorus detection is illustrated in Figure 45 on the next page.

8.6.3 Photoionization Detector

The PID uses an ultraviolet photon source to ionize aromatic and unsaturated compounds while giving little response to hydrocarbons or halocarbons. The resulting electron current is collected and amplified. It is used primarily for benzene, toluene, ethylbenzene, and xylene (BTEX) determination in petroleum analysis; and for these compounds plus unsaturates in environmental applications. PIDs have detectabilities comparable to FIDs. The photon energy from the ultraviolet lamp determines PID response: An 11.2–eV lamp will ionize com-

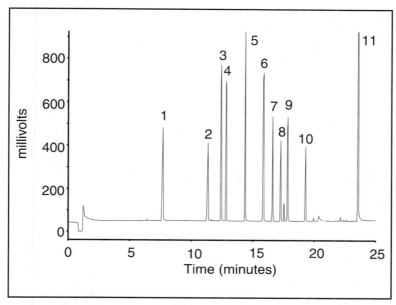

*Figure 45. Analylsis of drugs on a NPD. **Column**: 30-m x 0.25-mm i.d. x 0.25-μm film methylsilicone on fused silica. **Conditions**: oven, 90 °C for 1 min, 10 °C/min to 180 °C, 5 °C/min to 240 °C, 12 °C/min to 300 °C; helium carrier, 1 mL/min. **Peak identification**: (1) barbital; (2) secobarbital; (3) diphenhydramine; (4) PCP; (5) chlorpheneramine; (6) bromopheneramine; (7) methadone; (8) cocaine; (9) doxipine; (10) benztropine; (11) amoxipine.*

pounds with first ionization potentials equal to or less than 11.2 eV. Higher energy lamps are used for some organometallic compounds. Most organic compounds will respond to photons above 14 eV.

8.6.4 Electrolytic-Conductivity Detector

The ElCD is found almost exclusively in environmental water analyses. It responds to halogenated compounds in its *reductive* mode. Column eluent is heated to about 800 °C in a nickel catalyst reactor tube in the presence of hydrogen. Halogenated compounds produce HX (X = Cl, Fl, Br) which is entrained into a propanol solution where it increases the solution's conductivity, causing an output signal increase. The analysis of volatile hydrocarbons and halocarbons in water

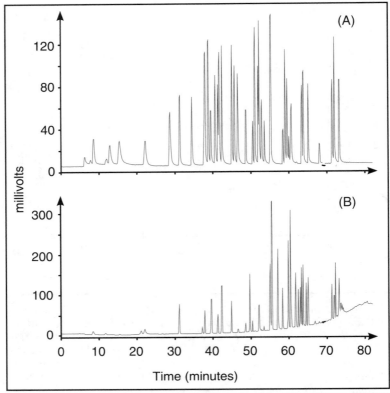

*Figure 46. Analysis of volatile water pollutants using (A) an ElCD and (B) a PID detector in series. **Column:** 105-m x 0.53-mm i.d. x 3-μm film Rtx 502.2. **Conditions:** oven, 35 °C, 10 min, 4 °C/min to 220 °C; helium carrrier 10 mL/min; ElCD, 250 °C, range 1x4; PID, 250 °C, range 1x64. **Sample:** 0.2 ppb each component in water.*

using an ElCD and PID detector in series is illustrated in Figure 46. The ElCD also operates in several *oxidative* modes, responding to sulfur or nitrogen containing organics.

8.6.5 Flame-Photometric Detector

The FPD responds to flame emission at specific wavelengths characteristic of phosphorous, sulfur, or tin heteroatoms. A hydrogen-rich flame burns the analytes as they come from the column. The spectral emissions are filtered with a narrow-bandpass interference filter, and a photomultiplier tube is illuminated with photons of the selected wavelength. FPD response to sulfur is proportional to the square of the sulfur concentration; many gas chromatographs include special circuitry or software to take the square root of the detector response and provide a linearized output signal. Hydrocarbon coelution can severely quench FPD response to heteroatoms, and sample cleanup may be required if large amounts of hydrocarbon are present in a sample. The FPD is applied in petroleum sulfur analysis, pesticide studies, and for some drug analyses.

8.6.6 Tandem Detection

Many GC detectors can also operate simultaneously, providing more than one kind of information about eluting solutes. In *series* operation, the column is connected to one detector and its exhaust is connected to another. This scheme works if the first detector does not destroy the sample components as they pass through. A PID, for example, does not break solutes down and is used frequently in series with an ElCD (as shown in Figure 46). In *parallel* operation, the column effluent is split into two streams which pass to two detectors. Now, a destructive detector such as a FID can be used with another such as a NPD. The column effluent is not always split 50:50. An uneven split may be advantageous with very sensitive detectors such as the ECD, since it ensures the transfer of an appropriately low fraction of high-concentration solutes.

✻ ✻ ✻

In this short chapter, we have summarized the major detectors used in open-tubular gas chromatography today. General purpose detectors such as the FID and TCD are applied in the majority of cases, while specific detectors are used when the peaks of selected compounds must be distinguished. Specific detection is not a substitute for good peak separation. It is an adjunct technique that can enhance the information from a chromatogram and provide confirmatory evidence for peak identity.

Part IX

Literature on Open-Tubular Columns

In this chapter, we list the combined references for this book, followed by a list of general references on open-tubular chromatography including textbooks; journals; Proceedings of the Symposia on Capillary Chromatography; and a list of publishers.

9.1 References

[1] D.H. Desty, J.N. Haresnape, and B.H.F. Whyman (to British Petroleum), *British Patent* 899,909 (application: April 9, 1959; issued: June 27, 1962).

[2] L.S. Ettre, in W.G. Jennings (ed.), *Applications of Glass Capillary Gas Chromatography,* M. Dekker, Inc., New York, 1981; pp. 1–47.

[3] L.S. Ettre, *Chromatographia* **16**, 18–25 (1982); **17**, 117 (1983).

[4] L.S. Ettre, *Anal. Chem.* **57**, 1419A–1438A (1985).

[5] L.S. Ettre, *Chromatographia* **34**, 513–528 (1992).

[6] L.S. Ettre, *HRC/CC* **10**, 221–230 (1987).

[7] M.J.E. Golay, in V.J. Coates, H.J. Noebels, and I.S. Fagerson (eds.), *Gas Chromatography (1957 Lansing Symposium),* Academic Press, New York, 1958; pp. 1–13.

[8] M.J.E. Golay, in D.H. Desty (ed.), *Gas Chromatography 1958 (Amsterdam Symposium),* Butterworths, London, 1958; pp. 36–55; pp.62–68.

[9] I.G. McWilliam and R.A. Dewar, *Nature (London)* **181**, 760 (1958).

[10] I.G. McWilliam and R.A. Dewar, in D.H. Desty, ed., *Gas Chromatography 1958 (Amsterdam Symposium),* Butterworths, London, 1958; pp. 142–152.

[11] R.D. Condon, *Anal. Chem.* **31**, 1717–1722 (1959).

[12] D.H. Desty, *Abh. Deut. Akad. Wiss, Berlin, Kl. Chem. Geol. Biol.* No.9 (1959); pp. 176–184.

[13] J.E. Lovelock, *J. Chromatogr.* **1**, 35–46 (1958).

[14] J.E. Lovelock, *Nature (London),* **182**, 663–664 (1958).

[15] A. Zlatkis and J.E. Lovelock, *Anal. Chem.* **31**, 620–621 (1959).

[16] S.R. Lipsky, R.A. Landowne, and J.E. Lovelock, *Anal. Chem.* **31**, 852–856 (1959).

[17] M.J.E. Golay, in R.P.W. Scott (ed.), *Gas Chromatography 1960 (Edinburgh Symposium),* Butterworths, London, 1960; pp. 139–143.

[18] I.U.P.A.C. Nomenclature for Chromatography, *Pure & Appl. Chem.* **65**, 819–872 (1993).

[19] D.H. Desty, J.N. Haresnape, and B.H.F. Whyman, *Anal. Chem.* **32**, 302–304 (1960).

[20] D.H. Desty, *Chromatographia* **8**, 452–455 (1975).

[21] R.D. Dandeneau and E.H. Zerenner, *HRC/CC* **2**, 351–356 (1979).

[22] L.S. Ettre, E.W. Cieplinski, and W. Averill, *J. Gas Chromatogr.* **1** (2), 7–16 (1963).

[23] E.R. Quiram, *Anal. Chem.* **35**, 593–595 (1963).

[24] L.S. Ettre and J.V. Hinshaw, *Basic Relationships of Gas Chromatography*. Advanstar Communications, 1993, 177 pp.

[25] G. Guichon, *Chromatogr. Rev.* **8**, 1–47 (1966).

[26] A.T. James and A.J.P. Martin, *Biochem. J.* **50**, 679–690 (1952).

[27] L.S. Ettre, *Chromatographia* **18**, 243–248 (1984).

[28] E. Kováts, *Helv. Chim. Acta* **41**, 1915–1932 (1958).

[29] E. Kováts, in J.C. Giddings, and R. A. Keller (eds.), *Advances in Chromatography Vol 1*, M. Dekker, Inc., New York, 1965; pp. 229–247.

[30] L. Rohrschneider, *J. Chromatogr.* **17**, 1–12 (1966).

[31] L. Rohrschneider, *J. Chromatogr.* **22**, 6–22 (1966).

[32] W. O. McReynolds, *J. Chromatogr. Sci.* **8**, 685–691 (1970).

[33] H. van den Dool and P.D. Kratz, *J. Chromatogr.* **11**, 463–471 (1963).

[34] L.S. Ettre, in A.B. Littlewood (ed.), *Gas Chromatography 1966 (Rome Symposium)*, Inst. of Petroleum, London, 1967; pp. 115–118.

[35] L. S. Ettre, *Chromatographia* **8**, 291–299 (1975).

[36] R. Kaiser, *Z. Anal. Chem.* **185**, 1–14 (1962).

[37] Reprinted with permission of Quadrex Corporation, New Haven, CT.

[38] Reprinted with permission of Restek Corporation, Bellefonte, PA.

[39] U.S.E.P.A Method 502.2, EPA-600/4-88/039, Dec. 1988.

[40] I. Halász and C. Horváth, *Anal. Chem.* **35**, 499–505 (1963).

[41] L.S. Ettre and J.E. Purcell, in J.C. Giddings and R.A. Keller (eds.), *Advances in Chromatography Vol. 10*, M. Dekker, Inc., New York, 1974; pp. 1–94.

[42] M. Mohnke and W. Saffert, in M. Van Swaay (ed.), *Gas Chromatography 1962 (Hamburg Symposium)*, Butterworths, London, 1962; pp. 216–224.

[43] I. Halász and C. Horváth, *Nature (London)* **197**, 71–72 (1963).

[44] J. E. Purcell, *Nature (London)* **201**, 1321–1322 (1964).

[45] W. Bertsch, F. Shunbo, R.C. Chang, and A. Zlatkis, *Chromatographia* **7**, 128–134 (1974).

[46] M. Novotny and A. Zlatkis, *Chromatogr. Rev.* **14**, 1–44 (1971).

[47] G. Alexander, *Chromatographia* **13**, 651–660 (1980).

[48] M. Novotny and K. Tesarik, *Chromatographia* **1**, 332–333 (1968).

[49] G. Alexander and G. A. F. M. Rutten, *J. Chromatogr.* **99**, 81–101 (1974).

[50] J. J. Franken, G. A. F. M. Rutten, and J. A. Rijks, *J. Chromatogr.* **126**, 117–132 (1976).

[51] F. Onuska and M. E. Comba, *J. Chromatogr.* **126**, 133–145 (1976).

[52] P. Sandra, M. Verstappe, and M. Verzele, *Chromatographia* **11**, 223–226 (1978).

[53] G. A. F. M. Rutten and J. A. Luyten, *J. Chromatogr.* **74**, 177–182 (1972).

[54] K. Grob and G. Grob, *J. Chromatogr.* **125**, 471–485 (1976).

[55] G. Schomburg, H. Husmann, and F. Weeke, *Chromatographia* **10**, 580–587 (1977).

[56] R. D. Dandenau and E. H. Zerenner, *HRC/CC* **2**, 351–356 (1979).

[57] K. L. Ogan, C. Reese, and R. P. W. Scott, *J. Chromatogr. Sci.* **20**, 425–428 (1982).

[58] G. Dijkstra and J. DeGoey, in D. H. Desty (ed.), *Gas Chromatography 1958 (Amsterdam Symposium)*, Butterworths, London, 1958; pp. 56–68.

[59] R. Kaiser, *Chromatographie in der Gasphase, II. Kapillar-Chromatographie*. Bibliographisches Institut, Mannheim, 1960; p. 67.

[60] G. Schomburg, H. Husmann, and F. Weeke, *J. Chromatogr.* **99**, 63–79 (1974).

[61] G. Schomburg and H. Husmann, *Chromatographia* **8**, 517–530 (1975).

[62] J. Bouche and M. Verzele, *J. Gas Chromatogr.* **6**, 501–505 (1968).

[63] E. L. Ilkova and E. A. Mistryukov, *J. Chromatogr. Sci.* **9**, 569–570 (1971).

[64] J. V. Hinshaw Jr., *J. Chromatogr. Sci.* **25**, 49–55 (1987). Reproduced by permission of Preston Publications, a division of Preston Industries, Inc.

[65] C. Madani, E. M. Chambaz, M. Rigaud, J. Durand, and P. Chebroux, *J. Chromatogr.* **126**, 161–160 (1976).

[66] L. Blomberg, J. Buijten, J. Gawdzick, and T. Wänmann, *Chromatographia* **11**, 521–525 (1978).

[67] L. Blomberg, J. Buijten, K. Markides, and T. Wänmann, *HRC/CC* **4**, 578–579 (1981).

[68] K. Grob, G. Grob, and K. Grob Jr., *J. Chromatogr.* **211**, 243–246 (1981).

[69] S. R. Lipsky and W. J. McMurray, *J. Chromatogr.* **239**, 61–69 (1982).

[70] K. Grob and G. Grob, *Chromatographia* **4**, 422–424 (1971).

[71] K. Grob Jr, G. Grob, and K. Grob, *J. Chromatogr.* **156**, 1–20 (1978).

[72] L.S. Ettre, *Chromatographia* **18**, 477–488 (1984).

[73] L.S. Ettre, *HRC/CC* **8**, 497–503 (1985).

[74] L. S. Ettre, in F. Bruner (ed.), *The Science of Chromatography*, Elsevier, Amsterdam, 1985; pp. 87–109.

[75] W. Seferovic, J. V. Hinshaw Jr., and L. S Ettre, *J. Chromatogr. Sci.* **24**, 374–382 (1986). Reproduced by permission of Preston Publications, a division of Preston Industries, Inc.

[76] J. J. van Deemter, F. J. Zuiderweg, and A. Klinkenberg, *Chem. Eng. Sci.* **5**, 271–289 (1956).

[77] J. V. Hinshaw, *LC/GC* **10**, 851–855 (1992).

[78] K. J. Hyver (ed.), *High-Resolution Chromatography*. Hewlett-Packard Co., Avondale, PA, 3rd ed., 1989; p. 1–16. Reproduced by permission of Hewlett-Packard Co.

[79] L. S. Ettre and E. W. March, *J. Chromatogr.* **91**, 5–24 (1974).

[80] L. S. Ettre, *Introduction to Open-Tubular Columns*. The Perkin-Elmer Corp., Norwalk, CT; 2nd ed., 1978; pp. 21–22.

[81] W. E. Harris and H. W. Habgood, *Programmed-Temperature Gas Chromatography*. Wiley, New York, 1966; 305 pp.

[82] N. G. Johansen, L. S. Ettre, and R. L Miller, *J. Chromatogr.* **256**, 393–417 (1983).

[83] N. H. Snow and H. M. McNair, *J. Chrom. Sci.* **30**, 271–275 (1992).

[84] L. S. Ettre, L. Mázor, and J. Takács, in J. C. Giddings and R. A. Keller (eds.), *Advances in Gas Chromatography, Vol. 8*, M. Dekker, Inc., New York, 1969; pp. 271–325.

[85] K. J. Klein, P. A. Larson, and J. A Breckenridge, *HRC/CC* **15**, 615–619 (1992).

[86] P. Sandra (ed.), *Sample Introduction in Capillary Gas Chromatography*, Huethig, Heidelberg, 1985; 265 pp.

[87] K. Grob, *Classical Split and Splitless Injection in Capillary Gas Chromatography*, Huethig, Heidelberg, 1986, 3rd ed., 1993; 571 pp.

[88] K. Grob, *On-Column Injection in Capillary Gas Chromatography*, Huethig, Heidelberg, 1987, 2nd printing, 1991; 591 pp.

[89] L. S. Ettre and W. Averill, *Anal. Chem.* **35**, 680–684 (1961).

[90] R. D. Condon and L. S. Ettre, in J. Krugers (ed.), *Instrumentation in Gas Chromatography*, Centrex Publishing Co., Eindhoven, The Netherlands, 1968; pp. 87–105.

[91] J. C. Sternberg, in J. C. Giddings and R. A Keller (eds.), *Advances in Chromatography, Vol. 2*, M. Dekker, Inc., New York, 1966; pp. 205–270.

[92] I. Halász and W. Schneider, *Anal. Chem.* **33**, 979–982 (1961).

[93] I. Halász and W. Schneider, in N. Brenner, J. E. Callen, and M. D. Weiss (eds.), *Gas Chromatography (1961 Lansing Symposium)*, Academic Press, New York, 1962; pp. 287–306.

[94] D. Jentzsch and W. Hövermann, *German Patent* 1,181,449 (October 20, 1962); *U.S. Patent* 3,257,775 (June 28, 1966).

[95] K. Grob and G. Grob, *Chromatographia* **5**, 3–12 (1972).

[96] G. Schomburg, H. Behlau, R. Dielmann, F. Weeke, and H. Husmann, *J. Chromatogr.* **142**, 87–102 (1977).

[97] K. Grob and K. Grob Jr., *J. Chromatogr.* **94**, 53–64 (1974).

[98] K. Grob and K. Grob Jr., *HRC/CC* **1**, 57–64 (1978).

[99] K. Grob Jr., *J. Chromatogr.* **237**, 15– 23(1982).

[100] K. Grob Jr. and R. Müller, *J. Chromatogr.* **244**, 185–196 (1982).

[101] M. Galli, S. Trestianu, and K. Grob Jr., *HRC/CC* **2**, 366–370 (1979).

[102] W. Vogt, K. Jacob, A. B. Ohnesorge, and H.W. Obwexer, *J. Chromatogr.* **186**, 197– (1979).

[103] F. Poy, S. Visani, and F. Terrosi, *J. Chromatogr.* **217**, 81–90 (1981).

[104] J. V. Hinshaw Jr., and W. Seferovic, in P. Sandra and W. Bertsch (eds.), *Sixth International Symposium on Capillary Chromatography (May 14–16, 1985, Riva del Garda)*, Huethig, Heidelberg, FRG, 1985; pp. 213–225.

[105] J. V. Hinshaw Jr. and W. Seferovic, *HRC/CC* **9**, 731–736 (1986).

[106] J. V. Hinshaw and L. S. Ettre, *HRC/CC* **12**, 251–254 (1989).

[107] M. Dressler, *Selective Gas Chromatographic Detectors*. Elsevier, Amsterdam, 1986; 192 pp.

[108] H. H. Hill and D. G. McMinn (eds.), *Detectors for Capillary Gas Chromatography*, Wiley, New York, 1992; 444 pp.

9.2 General References

The preceding reference list only includes direct references quoted in the text of our book and does not intend to provide an exhaustive listing of papers dealing with various aspects of open-tubular column gas chromatography. Today, it would be practically impossible to prepare such a compilation, due to the vast number of publications in this field. Therefore, here, at the end of our book, we would like to present only some guidelines about the available literature.

Additional, more detailed discussion of the theoretical and practical aspects and applications of open-tubular columns and of the GC systems designed for such columns can be found in three sources: *(1)* textbooks on open-tubular columns, *(2)* publications in journals dealing with chromatography, and *(3)* symposia proceedings.

9.2.1 Textbooks

Below, we list 19 books dealing specifically with open-tubular columns; the full names and addresses of the publishers are given in Section 9.2.4. The first book — which is now out-of-print — presented the first concise summary of the theory and practice of open-tubular column gas chromatography and its discussion is still fully valid. However, obviously, it did not deal with glass and fused-silica columns and with the more recent stationary phases. A special feature of this book was a complete listing of all publications on this subject up to July 1968, and it contained 288 references.

L.S. ETTRE: *Open-Tubular Columns in Gas Chromatography.* Plenum, 1965, 164 pp.

K. GROB: *Classical Split and Splitless Injection in Capillary Gas Chromatography.* Huethig, 1986; 324 pp.

K. GROB: *Making and Manipulating Capillary Columns for Gas Chromatography.* Huethig, 1986; 250 pp.

K. GROB: *On-Column Injection in Capillary Gas Chromatography.* Huethig, 1987, 2nd printing 1991; 591 pp.

W. HERRES: *Capillary Gas Chromatography - Fourier Transform Infrared Spectroscopy.* Huethig, 1987; 210 pp.

H.H. HILL and D.G. McMINN (eds.): *Detectors for Capillary Chromatography.* Wiley, 1992; 444 pp.

H. JAEGER: *Glass Capillary Chromatography in Clinical Medicine and Pharmacology.* Dekker, 1985; 656 pp.

W. JENNINGS: *Gas Chromatography with Glass Capillary Columns.* Academic Press, 1st ed. 1978, 184 pp; 2nd ed. 1980, 320 pp.

W.G. JENNINGS: *Comparison of Fused-Silica and Other Glass Columns in Gas Chromatography.* Huethig, 1981; 81 pp.

W. JENNINGS (ed.): *Applications of Glass Capillary Gas Chromatography.* Dekker, 1981; 629 pp.

W.G. JENNINGS and J.G. NIKELLY (eds.): *Capillary Chromatography — The Applications.* Huethig, 1991; 153 pp.

W. JENNINGS and T. SHIBAMOTO: *Quantitative Analysis of Flavor and Fragrance Volatiles by Glass Capillary Gas Chromatography.* Academic Press, 1980; 472 pp.

W.A. KÖNIG: *The Practice of Enantiomer Separation by Capillary Gas Chromatography.* Huethig, 1987; 120 pp.

M.L. LEE, F.J. YANG, and K.D. BARTLE: *Open-Tubular Column Gas Chromatography: Theory and Practice.* Wiley, 1984; 445 pp.

F.L. ONUSKA and F.W. KARASEK: *Open-Tubular Gas Chromatography in Environmental Sciences.* Plenum, 1984; 296 pp.

D. ROOD: *A Practical Guide to the Care, Maintenance and Troubleshooting of Capillary Gas Chromatography Systems.* Huethig, 1991; 191 pp.

P. SANDRA (ed.): *Sample Introduction in Capillary Gas Chromatography.* Huethig, 1985; 265 pp.

P. SANDRA and C. BICCHI (eds.): *Capillary Gas Chromatography in Essential Oil Analysis.* Huethig, 1987; 435 pp.

A. VAN ES: *High Speed, Narrow-Bore Capillary Gas Chromatography.* Huethig, 1992; 143 pp.

9.2.2 Chromatography Journals

There are seven journals dealing specially with chromatography (the publisher's name is in parentheses):

Chromatographia (Vieweg)

Journal of Chromatographic Science (Preston)

Journal of Chromatography (Elsevier)

Journal of High-Resolution Chromatography (Huethig)

Journal of Liquid Chromatography (Dekker)

Journal of Microcolumn Separations (Lee)

LC•GC Magazine (Advanstar)

To these, the general analytical-chemistry journals (such as e.g., *Analytical Chemistry,* published by the American Chemical Society) should also be added.

Except for the *Journal of Liquid Chromatography,* all these journals regularly publish papers on gas chromatography and today, at least 80 % of such papers report on work utilizing open-tubular columns.

9.2.3 Symposia Proceedings

A special characteristic of chromatography is the regular organization of annual or biannual international symposia at which reports on the newest development in theory, columns, system and applications have been presented. The two most important continuous general symposia series were the *International Chromatography Symposia* (ICS) organized by the (British) Chromatography Discussion Group (since 1982: The Chromatography Society*) and the Symposia on Advances in Chromatography* (SAC) organized by Professor Albert Zlatkis of the University of Houston. Originally, both series dealt with gas chromatogra-

No.	Date	Place	Proceedings			
			Editor	Publisher	Pages	Papers
1.	1975 May 4–7	Hindelang	R.E. Kaiser	Inst. of Chr.	303	16
2.	1977 May 1–5	Hindelang	R.E. Kaiser	Inst. of Chr.	425	25
3.	1979 April 29– May 3	Hindelang	R.E. Kaiser	Inst. of Chr.	580	32
4.	1981 May 3–7	Hindelang	R.E. Kaiser	Inst. of Chr.	902	45
5.	1983 April 26–28	Riva del Garda	J. Rijks	Elsevier	909	93
6.	1985 May 14–16	Riva del Garda	P. Sandra	Huethig	963	116
7.	1986 May 11–14	Gifu, Japan	D. Ishii, K. Jinno, P. Sandra	Nagoya	907	116
8.	1987 May 19–21	Riva del Garda	P. Sandra	Huethig	1272	147
9.	1988 May 16–19	Monterey, CA, USA	P. Sandra	Huethig	753	93
10.	1989 May 22–25	Riva del Garda	P. Sandra, G. Redant	Huethig	1616	192
11.	1990 May 14–17	Monterey, CA, USA	P. Sandra, G. Redant	Huethig	921	124
12.	1990 Sept. 11–14	Kobe, Japan	K. Jinno	Industr. Publ.	896	118
13.	1991 May 13–16	Riva del Garda	P. Sandra	Huethig	1710	262
14.	1992 May 25–29	Baltimore, MD, USA	P. Sandra, M.L. Lee	Found.	819	161
15.	1993 May 24–27	Riva del Garda	P. Sandra	Huethig	1744	235

Table 18. Symposia on Capillary Chromatography and the Proceedings of the Meetings. See the List of Publishers on page 175 for the full names and addresses of publishers.

phy only. They started to include papers on liquid chromatography (as an "associate technique") by the mid-1960s and changed their name to chromatography in general in 1969 (SAC) and 1974 (ICS), respectively. The ICS series started in 1956 (as a Symposium on "Vapour Phase Chromatography") and is a biannual event which is still active. The SAC series started in 1963 and the last was held in 1988 with a total of 25 symposia in this period. The proceedings of both symposia were regularly published, either in book form or as a separate volume of one of the international journals. Many of the fundamental papers on open-tubular columns were presented at these symposia.

In 1975, Dr. Rudolf E. Kaiser, the director of the Institute for Chromatography (Bad Dürkheim, Germany) organized in Hindelang, in the Bavarian Alps, a *Symposium on Glass Capillary Chromatography*. The meeting was an instant success; therefore, it was decided to make it a biannual event, a meeting series specially devoted to open-tubular (capillary) columns. Both the audience and the number of papers presented at these meetings grew very rapidly, and by 1983, it became clear that they had outgrown the facilities at Hindelang. In 1985, the symposium was moved to Northern Italy, to the resort of Riva on the northern tip of the *Lago di Garda*. Also, the meetings were made an annual event and their organization was taken over by a group headed by Professor P. Sandra of the University of Ghent (Belgium). From 1986 on, in addition to Riva del Garda, the meetings were periodically held in Japan and the United States. Recently, papers on capillary columns used in high-performance liquid chromatography and capillary zone electrophoresis have also been included in the program, but the bulk of the papers continue to deal with gas (and supercritical-fluid) chromatography. From the first meeting on, the proceedings of the symposium including the full text (or at least an extended abstract of most of the papers presented) had been available to the participants. Table 18 on page 173 lists the symposia of this series, with reference to the proceedings and indicating the number of papers included in it. As given, a total of 1777 papers were presented at these symposia since their beginning. These papers truly represent the newest advances in open-tubular column chromatography.

9.2.4 Publishers

Below we list the addresses of the publishers indicated previously:

Academic Press
 1250 Sixth Avenue
 San Diego, CA 92101-4311, USA

Advanstar Communications
 859 Willamette Street
 P.O. Box 10460
 Eugene, OR 97440-2460, USA

American Chemical Society
 1155 Sixteenth Street NW
 Washington, DC 20036, USA

Marcel Dekker, Inc.
 270 Madison Avenue
 New York, NY 10016, USA

Elsevier Science Publishers
 P.O. Box 330
 1000 AH Amsterdam, The Netherlands

Foundation for the International Symposium on
 Capillary Chromatography
 P.O. Box 164552
 Miami, FL, 33116-4552, USA

Dr. Alfred Huethig Verlag GmbH
 P.O. Box 102869
 D-69018 Heidelberg, Germany

Industrial Publishing & Consulting, Inc.
 Taiyo Building
 1-22-27 Hyakunin-cho, Shinjuku-ku
 Tokyo, 169, Japan

Institut für Chromatographie
P.O. Box 1141
D-67085 Bad Dürkheim, Germany

Dr. Milton L. Lee
Department of Chemistry
Brigham Young University
Provo, UT 84602, USA

The University of Nagoya Press
1 Furu-cho, Chikusa-ku
Nagoya 464, Japan

Plenum Publishing Corporation
233 Spring Street
New York, NY 10013-1578, USA

Preston Publications
7800 Merrimac Avenue
Niles, IL 60714-3498, USA

Vieweg Publishing Co.
P.O. Box 5829
D-65048 Wiesbaden 1, Germany

John Wiley & Sons
605 Third Avenue
New York, NY 10158-0012, USA

Part X

List of Symbols

In this book we have used the symbols and rules that are included in the I.U.P.A.C. Nomenclature for Chromatography[18].

10.1 List of Symbols Used

The symbols and acronyms used in this book are listed below, along with a short definition and the page number on which they first appear.

10.1.1 Symbols

A	"eddy diffusion" term in the van Deemter equation (93); true peak area (45)
A'	peak area calculated from peak width and height (45)
a_k	correction factor in sample capacity calculation (107)
B	longitudinal diffusion term in the van Deemter-Golay equation (50)

C	resistance to mass transfer term in the van Deemter–Golay equation (50)
C_M	mobile-phase resistance-to-mass-transfer term in the van Deemter–Golay equation (53)
C_S	stationary-phase resistance-to-mass-transfer term in the van Deemter–Golay equation (53)
c_L	concentration of the liquid phase in a coating solution [% wt/vol] (84)
c_n	number of carbon atoms in a molecule of a member of an homologous series (36)
d_c	column inner diameter (16)
d_f	liquid phase film thickness (16)
d_p	particle diameter of a stationary phase support (92)
D_M	gas diffusion coefficient in the mobile phase (52)
D_S	solute diffusion coefficient in the stationary phase (53)
F_a	mobile phase flow rate at column outlet, corrected to dry gas conditions (24)
F_c	mobile phase flow rate measured at column outlet, corrected to the column temperature (21)
F_{sp}	specific flow rate per gram stationary phase (123)
F_v	split vent flow (132)
$\overline{F}$	average column flow rate (123)
H	plate height [height equivalent to one theoretical plate; *HETP*] (47)
H_{meas}	measured plate height (60)
H_{min}	minimum plate height (51)
$H_{min\,(theor)}$	theoretical minimum plate height (54)
I	retention index (35)
I^T	linear [temperature-programmed] retention index (39)
j	carrier gas compression correction factor (22)
K	distribution constant [partition coefficient] (29)
k	retention factor (30)
k_M	retention factor of an unretained peak (32)
k_T	retention factor at temperature T (125)
L	column length (16)

List of Symbols

m_{max}	peak maximum height (45)
N	number of theoretical plates (47)
N_{eff}	effective plate number (49)
N/L	theoretical plates per meter (48)
P	relative pressure for a column (ratio of inlet and outlet pressures) (17)
Δp	pressure drop (17)
p_i	inlet pressure (17)
p_o	outlet pressure (17)
PN	peak number (64)
R_s	peak resolution (61)
r	relative retention of two peaks (33)
r_c	column inner radius (84)
r_T	temperature program rate (123)
s	split ratio (132)
SN	separation number (65)
t_0	initial isothermal period of a column temperature program (123)
t_M	unretained peak holdup time (19)
t_R	total retention time (28)
t_R'	adjusted retention time (28)
t_R^T	programmed-temperature retention time (39)
T_0	initial column temperature (123)
T_c	column temperature (24)
T_R	retention temperature (123)
T_t	column temperature at time t (123)
TZ	Trennzahl (65)
u	linear velocity (18)
$\bar{u}$	average carrier gas linear velocity (19)
$\bar{u}_{opt}$	optimum average carrier gas linear velocity (51)
$\bar{u}_{opt\,(theor)}$	theoretical optimum average carrier gas linear velocity (54)
u_i	linear velocity at the column inlet (19)
u_o	linear velocity at the column outlet (19)
V_{eff}	effective volume of a theoretical plate (107)
V_G	volume of mobile phase in the column (30)

V_{max}	maximum solvent vapor volume entering the column (107)
V_S	volume of stationary phase in the column (30)
w_b	peak width at base (43)
w_h	peak width at half height (43)
w_i	peak width at the inflection points (43)
W_S	weight of stationary phase in the column (124)
z	number of carbon atoms in the *n*-alkane eluting before the peak of interest (37)
$z + 1$	number of carbon atoms in the *n*-alkane eluting after the peak of interest (37)

10.1.2 Greek-Letter Symbols

α	separation factor of two consecutive peaks (32)
β	phase ratio of a column (30)
η	mobile phase viscosity (25)
φ	peak fidelity (130)
ρ	liquid stationary phase density (83)
σ	standard deviation of a Gaussian peak (44)
σ^2	variance of a Gaussian peak (130)
σ_c^2	peak variance due to the column, alone (130)
σ_e^2	peak variance due to extra-column contributions (130)
σ_{meas}^2	measured peak variance (130)
τ_d	detector response time constant (151)

10.2 Acronyms and Abbreviations

BHC	Hexachlorocyclohexane (156)
BTEX	Benzene, toluene, ethylbenzene, and xylenes (158)
DDD	Dichlorodiphenyl dichloroethane (156)
DDE	Dichlorodiphenyl dichloroethylene (156)
DDT	Dichlorodiphenyl trichloroethane (156)
ECD	Electron-capture detector (155)
ElCD	Electrolytic conductivity detector (159)

List of Symbols

FID	Flame-ionization detector (152)
FPD	Flame-photometric detector (160)
GC	Gas chromatography (13)
GLP	Good laboratory practice (86)
HETP	Height equivalent to one theoretical plate (47)
ITD	Ion-trap detector (157)
LC	Liquid chromatography (13)
MDQ	Minimum detectable quantity (150)
M.R.O.T.	Maximum recommended operating temperature (76)
MSD	Mass spectrometric detector (157)
PAH	Polynuclear aromatic hydrocarbon (40)
PCP	Pentachlorophenol (159)
PID	Photoionization detector (158)
PLOT	Porous-layer open-tubular [column] (78)
PTV	Programmed-temperature vaporizer (145)
SCOT	Support-coated open-tubular [column] (78)
SFC	Supercritical-fluid chromatography (13)
TCD	Thermal-conductivity detector (153)
TCEP	*tris*-Cyanoethoxypropane [stationary phase] (79)
U.S.E.P.A.	United States Environmental Protection Agency (74)
UTE%	Percent utilization of theoretical efficiency (60)
WCOT	Wall-coated open-tubular [column] (77)

Index

A

A term; 93, 95
Acid surface treatment; 80
Activity coefficient; 29
Adjusted retention time; 28
Adsorbent trap; 133
Adsorption
  in direct injection; 142
  in injection; 129
  in split injection; 136
  in splitless injection; 140
Adsorption column; 9, 153
Aldrin; 156
Alumina; 74 - 75
Aluminum cladding; 81
Amoxipine; 159
Argon-ionization detector; 4
Arochlors; 155
Aromatics; 158
Average column flow rate; 124
Average linear velocity
  definition; 19
  example; 21
  influence on speed and efficiency; 111
  measuring; 19

B

B term; 50
  definition; 52
Band broadening
  extra-column; 42, 79, 130, 154
  origins; 50
Bandwidth; 42
Barbital; 159
Base frequency; 155
Baseline; 11
Baseline resolution; 62
Benzene; 36, 39, 93, 158
Benztropine; 159
BHC; 156
Block copolymers; 84
Boiling point separation; 46
Bromopheneramine; 159
BTEX; 158
Bubble flow meter,; 21
Buffer volume; 133
1-Butanol; 39, 70

C

C term; 50, 78
  definition; 53
  example; 55
Capillary column
  nomenclature; 5
Carbon number; 36
Carbowax 20M; 70
Carbowax-1540; 8
Carrier gas; 15
  oxygen contamination; 157
  selection; 112
  viscosity; 24 - 26, 51, 126
  viscosity and column temperature; 126
  viscosity and pressure drop; 114
  with ECD; 157
  with TCD; 155
Cesium bead; 158

Cetyl alcohol; 8
4-Chloro-3-methylphenol; 72
2-Chlorophenol; 72
Chlorpheneramine; 159
Chromatograph, gas
   description; 10
Chromatography Society, The; 172
Coating efficiency; 60
Cocaine; 159
Coelution; 33, 120
Cold on-column injection; 142
Cold trapping
   in splitless injection; 138
Column
   definition; 10
   high temperature; 9
   surface activity; 76
Column coating; 82
Column deactivation; 78
Column efficiency; 46
Column etching; 78
Column flow; 21
   corrected; 24
   example; 23
Column heating
   See Oven
Column inlet linear velocity; 19
Column inlet pressure; 17
Column inner diameter; 7, 16
   effect on HETP; 99
Column installation
   fused-silica column; 81
   glass column; 80
Column length; 16, 19
   packed column; 95
   relationship to resolution; 103
   relationship to retention time; 103
Column material; 78
Column outlet linear velocity; 19
Column outlet pressure; 17
Column packing; 13
Column quality measurement; 86
Column temperature; 30, 51
   influence on distribution
      constant; 119
   influence on separation factor; 120
   isothermal; 119
   selection; 122
Column test mixture; 86
Compression correction factor; 22
Concentration-dependent detector;
   150
Concentric-tube splitter; 131
Constant mass flow; 26, 95, 126

Copper column; 79
Crude oil; 116
tris-Cyanoethoxypropane; 79

D

Data handling; 12
DDD; 156
DDE; 156
DDT; 156
Deactivation
   fused-silica column; 81
   glass column; 80
Decane; 70, 103
Decanoic acid, methyl ester; 70
Decomposition; 129
   in direct injection; 142
   in split injection; 136
   in splitless injection; 140
Desorption; 27
Desty, D.H.; 2, 4, 6
Detectivity; 150
Detector
   definition; 11
   requirements; 150
Detector selectivity; 150
Detector sensitivity; 150
Dewar, R.A.; 4
2,4-Dichlorophenol; 72
Dicyclohexylamine; 70
Didecyl phthalate; 75
2,6-Dimethylaniline; 69 - 70, 85
2,4-Dimethylphenol; 69 - 70, 72
2,4-Dinitrophenol; 72
Diphenhydramine; 159
Dipolar interactions; 68
Direct injection; 128, 140
   packed-column inlet; 141
Direct interface; 157
Distillation; 46
Distribution; 27
Distribution constant
   definition; 29
   influence of column
      temperature; 119
Dodecanoic acid, methyl ester; 70
Doxipine; 159
Drugs; 145, 158, 161
Dynamic coating method; 2, 82
Dynamic range; 150

E

Effective plates; 49
Efficiency
　percent utilization; 87
Electrolytic-conductivity
　detector; 159
Electron-capture detector; 137, 155
Electronic flow meter; 21
Elution; 27
Endosulfan; 156
Endrin; 156
Equilibrium concentration; 29
Equivalent chain length; 40
Ethylbenzene; 93, 158
2-Ethylhexanoic acid; 70

F

Fatty acid esters; 40
　unsaturated; 69
Film thickness; 16
　influence on phase ratio; 105
　influence on plate number; 87
　influence on sample capacity; 108
　selection; 106
Flame-ionization detector; 4, 137, 152
Flame-photometric detector; 160
Free-radical initiator; 84
Fused-silica column; 2, 6, 80
　external coating; 81
　flexible; 6, 81
　shape; 81
Future developments; 9

G

Gas holdup time; 19
Gas-gas diffusion coefficient; 52
Gas-solid adsorbent packing; 13
Gasoline; 101
Gauge pressure; 17
Gaussian peak shape; 42, 44
General-purpose detector; 150
Glass beads; 136
Glass column; 6, 79
Glass drawing machine; 6
Glass wool; 136
Golay equation
　See van Deemter–Golay equation
Golay, M.J.E.; 3, 50
Good laboratory practice; 86
Grob, K.; 6

H

Halocarbons; 155
Halogens; 159
Height equivalent to a theoretical
　plate; 47
Helium carrier gas; 112
Heptachlor; 156
Heptane; 36, 93, 106
Herbicides; 145, 158
Hexadecanol; 8
Hexane; 36, 93, 106
High-speed injection; 136
Hindelang Symposia; 174
Hot needle injection; 135
Hydrocarbons; 36
Hydrogen bonding; 68
Hydrogen carrier gas; 112
　and nickel tubing; 79
　with narrow-bore columns; 102

I

I.U.P.A.C.; 5
Inflection points; 43
Injection
　column requirements; 129
　development; 7
　non-split; 7
　syringe needle; 136
Injection volume
　in direct injection; 141
　in on-column injection; 143
　in split injection; 134
　in splitless injection; 137
Inlet flow rate; 131
　in direct injection; 140
Inlet overloading; 131
Inlet requirements; 131
Inlet selection; 147
Inlet temperature; 131, 140
International Chromatography
　Symposia; 172
Ion trap; 157
Isothermal oven; 10

J

j factor
　See Compression correction factor
Journals; 172

K

Kaiser, R.E.; 174
Kováts, E.; 36

L

Lindane; 155
Linear velocity; 30
　definition; 18
　effect on plate height; 50
Linearity
　injection; 128
　split injection; 134
Liner deactivation; 140
Liner packing; 136
　in direct injection; 142
　in PTV injection; 145
Lipsky, S.R.; 4
Liquid chromatography; 13, 15, 67
Longitudinal diffusion; 50, 52
Lovelock, J.E.; 4

M

Makeup gas; 151
　with ECD; 155
　with FID; 152
　with TCD; 154
Mass chromatogram; 157
Mass discrimination; 129
　in programmed-temperature
　　injection; 145 - 146
　in split injection; 135
Mass-flow dependent detector; 150
Mass-spectrometric detector; 157
　peak purity; 120
Maximum injection width; 130
Maximum recommended
　temperature; 76
McReynolds, W.O.; 39, 68
McWilliam, I.G.; 4
Mercury drop coating method; 83
Metal column; 6, 79
　surface activity; 8
Metallic impurities
　fused-silica column; 81
　glass column; 79
Methadone; 159
Methane peak time; 19
Method 502.2; 160
2-Methyl-2,6-dinitrophenol; 72
Minimum detectable quantity; 150
Minimum theoretical plate height
　definition; 51
　packed column; 94
Mobile phase; 13, 15
　definition; 10
Mobile phase volume; 30
Molecular sieves; 13, 78, 153

N

NaCl deposition; 80
Naphthalene; 69
Narrow-bore column; 99, 102
Nickel column; 79
Nitrogen carrier gas; 112
Nitrogen-phosphorus detector; 158
2-Nitrophenol; 69, 72
4-Nitrophenol; 72
Nomenclature; 5
Nonane; 93
Nononal; 70
Novotny, M.; 6

O

Octadecanoic acid, methyl ester; 69
Octane; 93
1-Octanol; 70
2-Octanone; 69, 103
On-column injection; 128, 142
　secondary cooling; 143
　temperature control; 143
Open split interface; 158
Open-tubular column
　evolution; 6
　nomenclature; 5
Optimum linear velocity; 19, 56
　definition; 51
　influence of carrier gas; 113
Orange oil; 12
Outlet linear velocity
　determination; 22
Oven
　air bath; 10
　conductive heating; 3
　isothermal; 10
　isothermal operation; 119
　size; 3
　temperature-programmed
　　operation; 122
Oxygen, in carrier gas; 76

P

Packed capillary column; 5
Packed column; 13
    compared to open-tubular; 92
    development; 3
    particle diameter; 92
Packed-column injection; 129
Paraffins; 36, 66, 68, 86
Partition coefficient
    See Distribution constant
Partitioning; 27
PCP; 159
Peak; 11
Peak dispersion; 44
Peak fidelity; 130
    in detector; 150
Peak maximum; 43, 45
Peak number; 64
Peak purity; 62
Peak shape; 42
Peak shape recovery
    in splitless injection; 138
Peak symmetry; 44
Peak tailing; 87
    detector; 151
Peak width; 43
    from area and height; 45
Peak width at base; 43
Peak width at half-height; 43
Peak width at inflection points; 43
Pentachlorophenol; 72
Pentane; 93, 106
Peroxide initiator; 84
Pesticides; 155, 158, 161
Phase ratio; 30
Phenol; 72
Phosphorus; 160
Photoionization detector; 158
Photomultiplier tube; 160
Phytane; 117
Plate height
    See Theoretical plate height
Plates
    See Effective plates
    See also Theoretical plates
Polyethylene; 85
Polyethylene glycol; 8, 69, 74
Polyimide; 81
Polynuclear aromatic hydrocarbons; 40, 106
Polywax-1000; 85
Porous polymer packing; 13

Porous-layer open-tubular column; 74, 78, 80
    definition; 14
Pressure conversion factors; 18
Pressure drop; 20
    definition; 17
    influence of carrier gas; 114
    packed column; 95
Pressure programming; 26, 52, 126
Programmed-temperature
    injection; 85, 144
Programmed-temperature retention
    time; 39
Publishers; 175
Pulse frequency; 155

Q

Quadrupole; 157
Quartz column; 2, 6, 80
Quartz injection liner; 137
Quartz wool; 137
Quenching; 157

R

Reference peaks
    in relative retention; 36
Relative pressure; 17
Relative retention; 33
Resistance-to-mass-transfer; 50, 53
Resolution; 42
    definition; 61
    example; 64
    relationship to column length; 103
Response time; 151
Retention factor
    definition; 30
    influence on plate height; 58
Retention gap
    in direct injection; 142
    in on-column injection; 144
    in splitless injection; 139
Retention index
    definition; 35
    example; 38
    in column quality control; 88
    isothermal; 36
    linear; 40
    other systems; 40
    programmed-temperature; 39
    stationary phase influence; 38
    stationary-phase classification; 68
Retention temperature; 123

Retention time; 28, 42
   influence of temperature; 119
   relationship to column length; 103
Riva del Garda Symposia; 174
Rohrschneider, L.; 39, 68
Rubidium bead; 158

S

Sample capacity; 129, 141
   influence of column
      parameters; 107
   measurement; 110
Sample vapor backflash; 131
Sample vaporization
   in programmed-temperature
      injection; 145
   in split injection; 133, 136
Sandra, P.; 174
Sandwich injection; 136
Secobarbital; 159
Secondary cooling; 143
Selective detector; 150, 155, 157
Separation factor
   definition; 32
   effect on resolution; 97
   influence of column temperature;
      120
   measurement; 32
Separation number; 65
Separation quality; 61
Septum purge; 134
Septumless on-column injection;
   144
Series operation of detectors; 161
Silicon dioxide; 80
Siloxane
   cyanopropyl; 71, 84
   phenylmethyl; 69
   vinyl; 71, 84
Solute bandwidth; 42
Solute retention; 28
Solute-stationary phase diffusion; 53
Solvent boiling point; 139
   in on-column injection; 142
Solvent effect
   in direct injection; 142
   in splitless injection; 138
Solvent flooding
   in on-column injection; 144
   in splitless injection; 139
Solvent management; 128
   in direct injection; 142
   in splitless injection; 137

Sorption; 27
Specific flow rate; 123
Speed of analysis; 96
   influence of carrier gas; 113
   influence of column diameter; 115
   influence of gas velocity; 112
Split injection; 128, 131
   development; 7
   origins; 2
Split point; 132
Split ratio; 131
   example; 133
Split vent flow; 132
Split vent programming; 137
Splitless injection; 128, 137
   origins; 2
Splitless sampling time; 137
Squalane; 8, 39, 75
Stainless steel; 6, 79
Standard deviation; 44
   peak; 47
Static coating method; 2, 83
   film thickness formula; 83
Stationary phase
   bleed; 8, 75
   bonded; 9, 14, 76, 85
   composition and polarity; 68, 71
   cross linking; 9, 14, 82, 84
   definition; 10
   film thickness; 8, 29, 32, 53
   high temperature; 85
   immobilized; 8, 14, 76
   influence on separation; 67
   maximum temperature; 74
   molecular weight; 84
   silicone; 9
   solute interaction; 68
   stability; 77, 82
Stationary-phase selection; 88
Stationary-phase volume; 30, 77
Stationary-phase classification; 68
Stationary-phase density; 124
Stationary-phase polarity; 68
Steric affinity; 68
Sulfur; 160
Supercritical-fluid chromatography;
   13, 15, 67
Support-coated open-tubular
   column; 78, 80
   definition; 14
Surface activity
   glass column; 79
   metal column; 79
Surface dehydration; 80

Surface silinization; 80
Surface tension
  glass column; 80
  stationary phase; 82
Surfactant; 79
Symbols; 177
  abbreviations; 180
  greek-letter; 180
Symposia on Advances in
  Chromatography; 172
Symposia on Capillary
  Chromatography; 174
Symposia proceedings; 172
Syringe needle
  in on-column injection; 144
Syringe needle fractionation; 135

T

Tangents to a peak; 43
Temperature programming; 10, 26, 39, 51, 66, 122
  computer simulation; 126
  multi-ramp; 123
  optimization; 123
Tesarik, K.; 6
Textbooks; 170
Theoretical plate height; 52
  minimum; 56
Theoretical plates
  calculating; 48
  description; 46
  effect of film thickness; 55
  measuring; 47, 87
Theoretical plates per meter; 48
Thermal-conductivity detector; 153
  micro; 4
Thermionic-specific detector; 158
Thick-film column; 2, 106
Thin-film column; 2, 106
Time constant; 151
Tin; 160
Toluene; 93, 158
Total ion chromatogram; 157
Total retention time; 28

Trace-level injection; 128, 140
Trennzahl; 65
2,4,6-Trichlorophenol; 72
Triglycerides; 69, 146

U

U.S.E.P.A.; 74
U.T.E.%
  definition; 60
  example; 60
Undecane; 70
Undecanoic acid, methyl ester; 70
Unretained peak time; 19
Unretained solute; 19
Unsaturates; 158

V

van Deemter equation; 93
van Deemter–Golay equation; 50
Vapor pressure
  solute; 29, 68
  stationary phase; 8
Variance; 130
Volatile hydrocarbons; 75
Volatile organic compounds; 160

W

Wall-coated open-tubular column; 77
  definition; 14
Wide-bore column; 8, 141, 144, 147

X

Xylene; 92 - 93, 158

Z

Zeolite; 74
Zlatkis, A.; 4, 172